TEXT BOOK
OF
MULTIPLE INTEGRALS

DPH MATHEMATICS SERIES

TEXT BOOK OF MULTIPLE INTEGRALS

By

A.K. Sharma

DPH

DISCOVERY PUBLISHING HOUSE

NEW DELHI-110002

Reprinted - 2019

First Published - 2005

ISBN: 978-81-7141-966-1

Text Book of Multiple Integrals

Published by:

DISCOVERY PUBLISHING HOUSE PVT. LTD.

4383/4B, Ansari Road, Darya Ganj

New Delhi-110 002 (India)

Phone: +91-11-23279245, 23253475; 43596065

E-mail: discoverybooksindia@gmail.com

discoverypublishinghouse@gmail.com

web: www.discoverypublishinggroup.com

Printed at:

Infinity Imaging Systems

Delhi

Preface

This book 'Text Book of Multiple Integrals' has been specially written to meet the requirement of B.Sc./B.A., students of various Indian Universities. The subject matter of this book has been discussed in such a simple way that the students find no difficulty to understand. Each chapter of this book contains complete theory and large number of solved example.

We hope this book will be found useful by the students and teachers on the various Indian Universities. We will appreciate any suggestion for the improvement of the book.

A.K. Sharma

CONTENTS

1

Multiple Integrals

(Double and Triple Integrals and Change of Order of Integration)

1.1 Double Integrals

The concept of double integral is an extension of the concept of a definite integral to the case of two arguments (*i.e.*, a two dimensional space). Let a function f (x, y) of the independent variables x and y be continuous inside some domain (region) A and on is boundary. Divide the domain A into n subdomains A_1, A_2,...An of areas δA_1, δA_2,...,δA_n. Let (x_r, y_r) be any point inside the rth elementary area δA_r. From the sum

$$S_n = f(x_1, y_1)\,\delta A_1 + f(x_2, y_2)\,\delta A_2 + \ldots + f(x_r, y_r)\,\delta A_r + \ldots + f(x_n, y_n)\,\delta A_n$$

$$= \sum_{r=1}^{n} f(x_r, y_r)\delta A_r. \qquad \ldots(1)$$

Now take the limit of the sum (1) as $n \to \infty$ in such a way that the largest of the areas δAr approaches to zero. This limit, if it exists, is called the *double integral* of the function f (x, y) over the domain A. It is denoted by $\iint_A f(x,y)$ dA and is real as "the double integral of f (x, y) over A".

Suppose the domain (region) A is divided into rectangular partitions by a network of lines parallel to the coordinate axes. Let dx be the length of area in Cartesian coordinated. The integral $\iint f(x,y)dA$ is written as $\iint_A f(x,y)\,dx\,dy$ and is called the double integral of f (x, y) over the region A.

1.2 Properties of a Double Integral

1. If the region A is partitioned into two parts, say A_1 and A_2, then

   $$\iint_A f(x,y)\;dx\,dy = \iint_{A_1} f(x,y)dx\,dy + \iint_{A_2} f(x,y)\;dx\,dy.$$

   Similarly, for a sub-division of A into three or more parts.

2. The double integral of the algebraic sum of a fixed number of functions is equal to the algebraic sum of the double integrals taken for each term. Thus

$$\iint_A [f_1(x,y)+f_2(x,y)+f_3(x,y)+\ldots]dx\,dy$$

$$= \iint_A f_1(x,y)dx\,dy + \iint_A f_2(x,y)dx\,dy + \iint_A f_3(x,y)dx\,dy + \ldots$$

3. A constant factor may be taken outside the integral sign. Thus

$$\iint_A mf(x,y)dx\,dy = m\iint_A f(x,y)dx\,dy,$$

where m is a constant.

1.3 Evaluation of Double Integrals

(a) If the region A be given by the inequalities $a \le x \le b$, $c \le y \le d$, then the double integral

$$\iint_A f(x,y)dx\,dy = \int_a^b\int_c^d f(x,y)dx\,dy$$

$$= \int_a^b\left[\int_c^d f(x,y)dy\right]dx, \qquad \ldots(1)$$

or

$$\iint_A f(x,y)dx\,dy = \int_c^d\int_a^b f(x,y)dy\,dx$$

$$= \int_c^d\left[\int_a^b f(x,y)dx\right]dy \qquad \ldots(2)$$

i.e., in this case the order of integration is immaterial, provided the limits of integration are changed accordingly.

Note: In formula (1) the definite integral $\int_c^d f(x,y)dy$ is calculated first. During this integration x is regarded as a constant. While in the formula (2) the definite integral $\int_a^b f(x,y)dx$ is calculated first and during this integration y is regarded as a constant.

(b) If the region A is bounded by the curves $y = f_1(x)$, $y = f_2(x)$, $x = a$ and $x = b$, then

$$\iint_A f(x,y)dx\,dy = \int_a^b\int_{f_1(x)}^{f_2(x)} f(x,y)dx\,dy$$

$$= \int_a^b\left[\int_{f_1(x)}^{f_2(x)} f(x,y)dy\right]dx,$$

where the integration with respect to y is performed first treating x as a constant.

Similarly, if the region A is bounded by the curves $x = f_1(y)$, $x = f_2(y)$, $y = c$, $y = d$, we have

$$\iint_A f(x,y)\,dx\,dy = \int_c^d \int_{f_1(y)}^{f_2(y)} f(x,y)\,dy\,dx$$

$$= \int_c^d \left[\int_{f_1(y)}^{f_2(y)} f(x,y)\,dx\right] dy,$$

where the integration with respect to x is performed first treating y as a constant.

Remember :

While evaluating double integrals, first integrate w.r.t. the variable haring variable limits (treating the other variable as constant) and then integrate w.r.t. the variable with constant limits.

Remark :

In the double integral $\int_a^b \int_c^d f(x,y)\,dx\,dy$, it is generally understood that the limits of integration c to d are those of y and the limits of integration a to b are those of x. However this is not a standard convention. Some authors regard these limits in the reverse order *i.e.* they regard the limits c to d as those of x and the limits a to b as those of y. So it is better to write this double integral as $\int_{x=a}^{b} \int_{y=c}^{d} f(x,y)\,dx\,dy$ so that there in no confusion about the limits. However in the double integral $\int_a^b \int_{f_1(x)}^{f_2(x)} F(x,y)\,dx\,dy$, there is no confusion about the limits. Obviously the variable limits are those of y because they are in terms of x and so the constant limits must be those of x. Here the first integration must be performed with respect to y regarding x as constant.

1.4 To Express a Double Integral in Terms of Polar Coordinates

Let a function $f(r, \theta)$ of the polar coordinates (r, θ) be continuous inside some region A and on its boundary. Let the region A be bounded by the curves $r = f_1(\theta)$, $r = f_2(\theta)$ and the lines $\theta = \theta_1$, $\theta = \theta_2$.

Divide the area A into elements by a series of concentric circular arcs with centre at origin and successive radii differing by equal amounts and a series of straight lines drawn through the origin at equal intervals of angles. Let δr be the distance between two consecutive circles and $\delta\theta$ be the angle between two consecutive lines. There is thus a network of elementary areas

(say n in number) of which a typical one is PQRS. If P is the point (r, θ), the area of the element PQRS situated at the point P is $\frac{1}{2}(r+\delta r)^2\,\delta\theta - \frac{1}{2}r^2\,\delta\theta$,
$= r\delta\,\theta\,\delta r$, by neglecting the term $\frac{1}{2}(\delta r)^2\,\delta\theta$ being an infinitesimal of higher order.

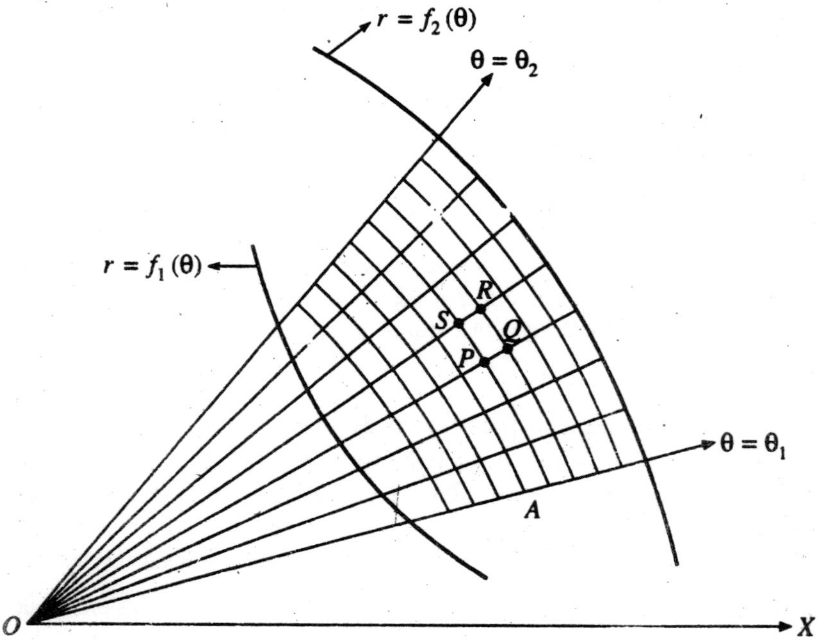

Fig. 1.1

Now by the definition of the double integers of f (r, θ) over the region A, we have

$$\iint_A f(r,\theta)dA = \lim_{\substack{n\to\infty \\ \delta r\to 0,\delta\theta\to 0}} \sum_{k=1}^{n} f(r_k,\theta_k)r_k\delta\theta\,\delta r,$$

where rk δq δr is the area of the element started at the point (r_k, θ_k).

Using the area of integration, this double integral is generally written as

$$\int_{\theta 1}^{\theta 2}\int_{f1(\theta)}^{f2(\theta)} f(r,\theta)d\theta\,dr\text{, or }\int_{\theta 1}^{\theta 2} d\theta\int_{f1(\theta)}^{f2(\theta)} f(r,\theta)dr.$$

The first integration is performed with respect to r, keeping θ as a constant. After substituting the limits for r, the second integration with respect to θ is performed.

Remark:

The area of the typical element PQRS situated at the point P (r, θ) can also be found as below :

We have OP = r, OQ = r + δr so that PQ = δr. Also PS is the arc of a circle of radius r subtending an angle δθ at the centre of the circle and so arc PS = r δθ. Therefore the area of the element PQRS is δr. r δθ *i.e.*, r δθ δr.

1.5 Triple Integrals

Let the function f(x, y, z) of the point P (x, y, z) be continuous for all points within a finite region V and on its boundary. Divide the region V into n parts; let $\delta V_1, \delta V_2, \ldots, \theta V_n$ be their volumes. Take a point in each part and form the sum

$$S_n = f(x_1, y_1, z_1)\,\delta V_1 + f(x_2, y_2, z_2)\,\delta V_2 + \ldots + f(x_n, y_n, z_n)\,\delta V_n$$

$$= \sum_{r=1}^{n} f(x_r, y_r, z_r)\,\delta V_r\,. \qquad \ldots(1)$$

Then the limit to which the sum (1) tends when n tends to infinity and the dimensions of each sub-division tend to zero, is called the triple integral of the function f(x, y, z) over the region V. This is denoted by

$$\iiint_V f(x,y,z)dV \text{ or } \iiint_V f(x,y,z)dx\,dy\,dz.$$

1.6 Evaluation of Triple Integrals

(a) If the region V be specified by the inequalities

$a \le x \le b,\ c \le y \le d,\ e \le z \le f,$

then the triple integral

$$\iiint_V f(x,y,z)dx\,dy\,dz$$

$$= \int_a^b\int_c^d\int_e^f f(x,y,z)dx\,dy\,dz$$

$$= \int_a^b dx\int_c^d dy\int_e^f f(x,y,z)\,dz.$$

Here the order of integration is immaterial and the integration with respect to any of x,y and z can be programed first.

(b) If the limits of z are given as functions of x and y, the limits of y as functions of x while x takes the constant values say from x = a to x = b, then

$$\iiint_V f(x,y,z)dx\,dy\,dz = \int_a^b dx\int_{y_1(x)}^{y_2(x)} dy\int_{z_1(x,y)}^{z_2(x,y)} f(x,y,z)dz.$$

The integration with respect to z is performed first regarding x and y as constants, then the integration w.r.t. y is performed regarding x as a constant and in the last we perform the integration w.r.t. x.

1.7 Change of Order of Integration

If in a double integral the limits of integration of both x and y are constant, we can generally integrate $\iint f(x,y)dx\,dy$ in either order. But if the limits of y are functions of x, we must first integrate w.r.t. y regarding x as constant and then integrate w.r.t. x. In this case the order of integration can be chaged only if we find the new limits of x as functions of y and the new constant of y.

1.8 Change of Variables in a Double Integral

Sometimes, the evaluation of a double integral becomes more convenient by a suitable change of variable from one system to another system.

Let the variables in the double integral $\iint_A f(x,y)dx\,dy$ be changed from x,y to u, v where x = ϕ (y, v) and y = ψ (u,v).

Then on substituting for x and y, the double integral is transformed to $\iint_{A'} F(u,v)$ J du dv, where J (u, v) is the Jacobian of x,y w.r.t. u,v *i.e.*,

$$J = \frac{\partial(x,y)}{\partial(u,v)} = \begin{vmatrix} \frac{\partial x}{\partial u} & \frac{\partial y}{\partial u} \\ \frac{\partial x}{\partial v} & \frac{\partial y}{\partial v} \end{vmatrix},$$

and A' is the region in the uv-plane corresponding to the region A in the xy-plane. Thus remember that dxdy = J dudv.

Special Case : *Change to polar coordinates from the Cartesian coordinates.*

To change the variables from cartesians to polar coordinates we put x = r cos θ, y = r sin θ. In this case

$$= J = \frac{\partial(x,y)}{\partial(r,\theta)} = \begin{vmatrix} \frac{\partial x}{\partial r} & \frac{\partial y}{\partial \theta} \\ \frac{\partial x}{\partial r} & \frac{\partial y}{\partial \theta} \end{vmatrix} = \begin{vmatrix} \cos\theta & -r\sin\theta \\ \sin\theta & r\cos\theta \end{vmatrix} = r,$$

and therefore dxdy = J dθ dr = r dθ dr.

This change is specially usefully useful when the region of integration is a circle or a part of a circle.

MISCELANEOUS EXAMPLES

Example 1:

Evaluate $\iint r^2 d\theta\, dr$ *over the area of the circle r = a cos θ.*

Solution:

The circle r = a cos θ passes through the pole and the diameter through the pole is initial line. The region of integration can be covered by radial strips originating from r = 0 and terminating at r = a cos θ. From the equation of the circle, we have r = 0 when cos θ = 0 *i.e.*, θ = ± π/2. Therefore for the given area θ varies from – π/2 to π/2. Therefore, the required integral

$$= \int_{\theta=-\pi/2}^{\pi/2}\int_{r=0}^{a\cos\theta} r^2 d\theta\, dr = \int_{-\pi/2}^{\pi/2}\left[\int_0^{a\cos\theta} r^2 dr\right]d\theta$$

$$= \int_{-\pi/2}^{\pi/2}\left[\frac{r^3}{3}\right]_0^{a\cos\theta} d\theta$$

$$= \int_{-\pi/2}^{\pi/2}\frac{a^3\cos^3\theta}{3}d\theta = \frac{2a^3}{3}\int_0^{\pi/2}\cos^3\theta\, d\theta = \frac{2a^3}{3}\cdot\frac{2}{3.1} = \frac{4a^3}{9}$$

Example 2:

Evaluate $\iint \frac{r\, d\theta\, dr}{\sqrt{(a^2+r^2)}}$ *over one loop of the lemniscate* $r^2 = a^2$ *cos 2θ.*

Solution:

In the equation of the lemniscate $r^2 = a^2\cos^2\theta$, putting r = 0, we get cos 2θ, putting r = 0, we get cos 2θ = 0 *i.e.*, 2θ = ± π/4. Therefore, one loop of the given lemniscate θ varies from –π/4 and r varies from 0 to a √(cos 2θ).

Therefore the required integral

$$= \int_{\theta=-\pi/4}^{\pi/4}\int_{r=0}^{a\sqrt{(\cos 2\theta)}} \frac{r\, d\theta\, dr}{\sqrt{(a^2+r^2)}}$$

$$= \int_{-\pi/4}^{\pi/4}\int_{r=0}^{a\sqrt{(\cos 2\theta)}} \frac{1}{2}(a^2+r^2)^{-1/2}(2r)d\theta\, dr$$

$$= \int_{-\pi/4}^{\pi/4}\left[(a^2+r^2)^{1/2}\right]_0^{a\sqrt{(\cos 2\theta)}} d\theta$$

$$= \int_{-\pi/4}^{\pi/4}[a(1+\cos 2\theta)^{1/2} - a]d\theta$$

$$= 2a\int_0^{\pi/4}[(2\cos^2\theta)^{1/2} - 1]d\theta = 2a\int_0^{\pi/4}(\sqrt{2}\cos\theta - 1)d\theta$$

$$= 2a\left[\sqrt{2}\sin\theta - \theta\right]_0^{\pi/4} = 2a\left[\sqrt{2}.\frac{1}{\sqrt{2}} - \frac{\pi}{4}\right]$$

$$= 2a\left[1 - \frac{\pi}{4}\right] = \frac{a}{2}(4 - \pi).$$

Example 3:

Find by double integration the area lying inside the cardioid $r = a\,(1 + \cos\theta)$ and outside the circle $r = a$.

Solution:

Eliminating r between the given equations of the cardioid $r = a\,(1 + \cos\theta)$ and the circle $r = a$, we have $a = a\,(1 + \cos\theta)$ or $\cos\theta = 0$ *i.e.*, $\theta = \pm\,\pi/2$.

Thus, the region of integration A is enclosed by $r = a$, $r = a\,(1 + \cos\theta)$, $\theta = -\,\pi/2$, $\theta = \pi/2$.

$\therefore$ the required area $= \iint_A r\,d\theta\,dr = \int_{-\pi/2}^{\pi/2}\int_a^{a(1+\cos\theta)} r\,d\theta\,dr$

$$= \int_{-\pi/2}^{\pi/2}\left[\frac{r^2}{2}\right]_a^{a(1+\cos\theta)} d\theta = \frac{1}{2}\int_{-\pi/2}^{\pi/2}[a^2(1+\cos\theta)^2 - a^2]d\theta$$

$$= \frac{a^2}{2}\int_{-\pi/2}^{\pi/2}(1 + \cos^2\theta + 2\cos\theta - 1)d\theta$$

$$= \frac{a^2}{2}.2\int_0^{\pi/2}[\cos^2\theta + 2\cos\theta]d\theta$$

$$= a^2\left[\frac{1}{2}.\frac{1}{2}\pi + 2\{\sin\theta\}_0^{\pi/2}\right] = a^2\left[\frac{1}{4}\pi + 2\right] = \frac{a^4}{4}(\pi + 8).$$

Example 4:

Find the mass of a loop of the lemniscate $r^2 = a^2\sin^2\theta$ if density $\rho = kr^2$.

Solution:

In the equation of the lemniscate $r^2 = a^2\sin^2\theta$, putting $r = 0$, we get $\sin 2\theta = 0$ *i.e.*, $2\theta = 0, \pi$ *i.e.*, $\theta = 0, \frac{1}{2}\pi$. Therefore, for one loop of the given lemniscate θ varies from 0 to $\pi/2$ and r varies from 0 to $a\sqrt{(\sin 2\theta)}$.

mass of a loop of the lemniscate

$$= \int_{\theta=0}^{\pi/2}\int_{r=0}^{a\sqrt{(\sin 2\theta)}} \rho r\,d\theta\,dr = \int_{\theta=0}^{\pi/2}\int_{r=0}^{a\sqrt{(\sin 2\theta)}} kr^2.r\,d\theta\,dr$$

$$= k\int_{\theta=0}^{\pi/2}\int_{r=0}^{a\sqrt{(\sin 2\theta)}} r^3 d\theta\, dr = k.\int_0^{\pi/2}\left[\frac{r^4}{4}\right]_{r=0}^{a\sqrt{(\sin 2\theta)}} d\theta$$

$$= \frac{ka^4}{4}\int_0^{\pi/2}\sin^2 2\theta\, d\theta = \frac{ka^4}{8}\int_0^{\pi/2}(1-\cos 4\theta)d\theta$$

$$= \frac{ka^4}{8}\left[\theta - \frac{\sin 4\theta}{4}\right]_0^{\pi/2} = \frac{ka^4}{8}.\frac{\pi}{2} = \frac{\pi ka^4}{16}$$

Example 5:

Integrate r sin θ over the area of the cardioid r = a (1+ cos θ) lying above the initial line.

Solution:

For the area of the cardioid r = a (1 + cos θ) above the initial line θ varies from 0 to π. Also for that required area r varies from r = 0 to r = a (1 + cos θ). If A denotes the region consisting of the area of the cardioid lying above the initial line, then the required integral

$$= \iint_A r\sin\theta\, dA = \int_0^{\pi}\int_0^{a(1+\cos\theta)} r\sin\theta\, r\, d\theta\, dr$$

$$= \int_0^{\pi}\sin\theta\left[\frac{r^3}{3}\right]_0^{a(1+\cos\theta)} d\theta = \frac{a^3}{3}\int_0^{\pi}\sin\theta(1+\cos\theta)^3 d\theta$$

$$= \frac{a^3}{3}\int_0^{\pi} 2\sin\frac{\theta}{2}\cos\frac{\theta}{2}\left(2\cos^2\frac{\theta}{2}\right)^3 d\theta$$

$$= \frac{16a^3}{3}\int_0^{\pi/2}\sin\phi\cos^7\phi.2d\phi, \text{ Puting } \frac{\theta}{2} = \phi \text{ so that } d\theta = 2d\phi$$

$$= 32.\frac{a^3}{3}\left[-\frac{\cos^8\phi}{8}\right]_0^{\pi/2} = \frac{32a^3}{3}\left[0+\frac{1}{8}\right] = \frac{4a^3}{3}.$$

Example 6:

Evaluate $\iint_R r^2\sin\theta\, d\theta\, dr$ *where R is the circle r = 2a cos θ.*

Solution:

The given integral $I = \int_{\theta=-\pi/2}^{\pi/2}\int_{r=0}^{2a\cos\theta} r^2\sin\theta\, d\theta\, dr$

$$= \int_{-\pi/2}^{\pi/2}\left[\int_{r=0}^{2a\cos\theta} r^2 dr\right]\sin\theta\, d\theta$$

$$= \int_{-\pi/2}^{\pi/2} \left[\frac{r^3}{3}\right]_0^{2a\cos\theta} \sin\theta\, d\theta$$

$$= \int_{-\pi/2}^{\pi/2} \frac{8a^3}{3} \cos^3\theta \sin\theta\, d\theta$$

$= 0$ because $\cos^3\theta \sin\theta$ is an odd function of θ.

$$\left[\text{Note that} \int_{-a}^{a} f(x)dx = 0 \text{ if } f(x) = -f(x)\right]$$

Example 7:

Find by double integration the area lying inside the circle $r = a \sin\theta$ and outside the cardioid $r = a (1 + \cos\theta)$.

Solution:

The given circle is $r = a\sin\theta$ and the cardioid is $r = a(1 - \cos\theta)$. Note that the given circle passes through the pole and the diameter through the pole makes an angle $\pi/2$ with the initial line. Eliminating r between the two equations, we have

$$a\sin\theta = a(1 - \cos\theta)$$

or $$1 = \frac{\sin\theta}{1-\cos\theta} = \frac{2\sin\frac{1}{2}\theta\cos\frac{1}{2}\theta}{2\cos^2\frac{1}{2}\theta} = \tan\frac{\theta}{2}$$

or $$\frac{1}{2}\theta = \frac{1}{4}\pi \quad i.e., \ \theta = \pi/2.$$

Thus, the two curves meet at the point where $\theta = \pi/2$. Also for both the curves $r = 0$ when $\theta = 0$ and so the two curves also meet at the pole O where θ 0. To cover the required area the limits of integration for r are $a(1 - \cos\theta)$ to $a\sin\theta$ and for θ are 0 to $\pi/2$. Therefore the required area

$$\int_0^{\pi/2}\int_{a(1-\cos\theta)}^{a\sin\theta} r\, d\theta\, dr = \int_0^{\pi/2}\left[\frac{r^2}{2}\right]_{a(1-\cos\theta)}^{a\sin\theta} d\theta$$

$$= \frac{1}{2}\int_0^{\pi/2}[a^2\sin^2\theta - a^2(1-\cos\theta)^2]d\theta$$

$$= \frac{a^2}{2}\int_0^{\pi/2}[\sin^2\theta - 1 + 2\cos\theta - \cos^2\theta]d\theta$$

$$= \frac{a^2}{2}\left[\frac{1}{2}\cdot\frac{\pi}{2} - \frac{\pi}{2} + 2.1 - \frac{1}{2}\cdot\frac{\pi}{2}\right] = \frac{a^2}{4}\left[1 - \frac{\pi}{2}\right] = \frac{a^2}{4}(4-\pi).$$

Example 8:

Transform the integral $\int_0^2\int_0^{\sqrt{(2x-x^2)}} \frac{x\,dx\,dy}{\sqrt{(x^2+y^2)}}$ *by changing to polar coordinates and hence evaluate it.*

Solution:

From the limits of integration it is obvious that the region of integration is bounded by $y = 0$, $y = \sqrt{(2x - x^2)}$ and $x = 0$, $x = 2$ *i.e.*, the region of integration is the area of the circle $x^2 + y^2 - 2x = 0$ between the lines $x = 0$, $x = 2$ and lying above the axis of x*i.e.*, the line $y = 0$,

Putting $x = r\cos\theta$, $y = r\sin\theta$ the corresponding polar equation of the circle is $r^2(\cos^2\theta + \sin^2\theta) - 2r\cos\theta = 0$, or $r = 2\cos\theta$.

From the figure it is obvious that r varies from 0 to $2\cos\theta$ and θ varies from 0 to $\pi/2$. Note that at the point A of the circle $\theta = 0$ and at the point O, $r = 0$ and so from $r = 2\cos\theta$, we get $\theta = \pi/2$ at O.

The polar equivalent of elementary area dxdy is $r\,d\theta\,dr$.

$\therefore \iint_A f(x,y)dx\,dy = \iint_A f(r\cos\theta, r\sin\theta)r\,d\theta\,dr$, where A is the region of integration.

Hence transforming to polar coordinates, the given double integral

$$= \int_\theta^{\pi/2}\int_{r=0}^{2\cos\theta} \frac{r\cos\theta}{r} r\,d\theta\,dr$$

$$= \int_\theta^{\pi/2} \cos\theta \left[\frac{r^2}{2}\right]_0^{2\cos\theta} d\theta$$

$$\int_\theta^{\pi/2} \frac{1}{2}\cos\theta.4\cos^2\theta d\theta = 2\int_0^{\pi/2} \cos^3 d\theta = 2.\frac{2}{3} = \frac{4}{3}.$$

Example 9:

Find by double integration the area lying inside the cardioid $r = 1 + \cos\theta$ *and outside the parabola* $r(1 + \cos\theta) = 1$.

Solution:

Eliminating r between the given equations of the cardioid and the parabola, we have $(1 + \cos\theta) = 1/(1 + \cos\theta)$ or $(1 + \cos\theta)^2 = 1$

or $\quad \cos^2\theta + 2\cos\theta = 0$ or $\cos\theta(2 + \cos\theta) = 0$

or $\quad \cos\theta = 0$, because $\cos\theta$ cannot be equal to -2

or $\quad \theta = \pm\pi/2$.

Thus, the two curves intersect at the point where $\theta = -\pi/2$ and $\theta = \pi/2$.

Therefore, the required area is enclosed by $r = 1/(1 + \cos\theta)$, $r = (1 + \cos\theta)$, $\theta = -\pi/2$, $\theta = \pi/2$.

Hence the required area

$$= \int_{-\pi/2}^{\pi/2}\int_{1/(1+\cos\theta)}^{(1+\cos\theta)} r\,d\theta\,dr = \int_{-\pi/2}^{\pi/2}\left[\frac{1}{2}r^2\right]_{1/(1+\cos\theta)}^{(1+\cos\theta)} d\theta$$

$$= \frac{1}{2}\int_{-\pi/2}^{\pi/2}\left[(1+\cos\theta)^2 - \frac{1}{(1+\cos\theta)^2}\right]d\theta$$

$$= 2\cdot\frac{1}{2}\int_0^{\pi/2}\left[(1+2\cos\theta+\cos^2\theta) - \frac{1}{(2\cos^2\frac{1}{2}\theta)^2}\right]d\theta$$

$$= \int_0^{\pi/2}(1+2\cos\theta)d\theta + \int_0^{\pi/2}\cos^2\theta\,d\theta - \frac{1}{4}\int_0^{\pi/2}\sec^4\frac{2}{2}\theta\,d\theta$$

$$= [\theta + 2\sin\theta]_0^{\pi/2} + \frac{1}{2}\cdot\frac{\pi}{2} - \frac{1}{4}\int_0^{\pi/2}(1+\tan^2\frac{1}{2}\theta)\sec^2\frac{1}{2}\theta\,d\theta$$

$$= \frac{\pi}{2} + 2 + \frac{\pi}{4} - \frac{1}{4}\int_0^{\pi/2}\left[\sec^2\frac{1}{2}\theta + 2(\tan^2\frac{1}{2}\theta)(\frac{1}{2}\sec^2\frac{1}{2}\theta)\right]d\theta$$

$$= \frac{3\pi}{4} + 2 - \frac{1}{4}\left[2\tan\frac{1}{2}\theta + \frac{2}{3}\tan^3\frac{1}{2}\theta\right]_0^{\pi/2}$$

$$= \frac{3\pi}{4} + 2 - \frac{1}{4}\left[2 + \frac{2}{3}\right] = \frac{3\pi}{4} + 2 - \frac{2}{3} = \frac{3\pi}{4} + \frac{4}{3} = \frac{(9\pi+16)}{12}.$$

Example 10:

Transform the following double integrals to polar coordinates and hence evaluate them.

(i) $\int_{y=0}^{a}\int_{x=0}^{\sqrt{(a^2-y^2)}}(a^2 - x^2 - y^2)\,dx\,dy.$

(ii) $\int_0^1\int_x^{\sqrt{(a^2-y^2)}}(x^2 + y^2)\,dx\,dy$

(iii) $\int_0^a\int_0^{\sqrt{(2a-x^2)}} y^2\sqrt{(x^2+y^2)}\,dx\,dy$

Solution:

(i) The given double integral $I = \int_{y=0}^{a}\int_{x=0}^{\sqrt{(a^2-y^2)}}[a^2 - (x^2+y^2)]\,dx\,dy.$

From the limits of integration it is obvious that the region of integration R is bounded by $x = 0$, $x = \sqrt{(a^2-y^2)}$ and $y = 0$, $y = a$.

Thus, the region of integration is the area OAB of the circle $x^2 + y^2 = a^2$ lying in the positive quadrant.

Putting $x = r\cos\theta$, $y = r\sin\theta$ the corresponding polar equation of the circle is $r = a$.

From the figure it is obvious that for the area OAB, r varies from 0 to a and θ varies from 0 to $\pi/2$. Also the polar equivalent of dx dy is $r\,d\theta\,dr$.

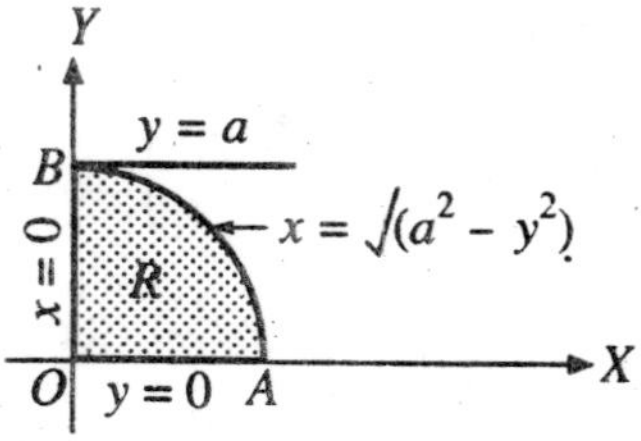

Fig. 1.2

$$\therefore I = \int_{\theta=0}^{\pi/2}\int_{r=0}^{a}(a^2 - r^2)r\,d\theta\,dr,$$

$$= \int_{\theta=0}^{\pi/2}\int_{r=0}^{a}[a^2r - r^3]d\theta\,dr = \int_0^{\pi/2}\left[\frac{a^2r^2}{2} - \frac{r^4}{4}\right]_{r=0}^{a} d\theta$$

$$= \int_0^{\pi/2}\left[\frac{a^4}{2} - \frac{a^4}{4}\right]d\theta = \frac{a^4}{4}\int_0^{\pi/2}d\theta = \frac{a^4}{4}[\theta]_0^{\pi/2} = \frac{\pi a^4}{8}.$$

(ii) The given double integral $I = \int_{x=0}^{1}\int_{y=x}^{\sqrt{(2x-x^2)}}(x^2 + y^2)dx\,dy.$

Here the region of integration R is bounded by $y = x$, $y = \sqrt{(2x-x^2)}$ and $x = 0$, $x = 1$ *i.e.*, the region of integration is the area OBCO of the circle $x^2 + y^2 - 2x = 0$ bounded by the lines $y = x$, $x = 0$ and $x = 1$.

Putting $x = r\cos\theta$, $y = r\sin\theta$ the corresponding polar equation of the circle is $r^2(\cos^2\theta + \sin^2\theta) - 2r\cos\theta = 0$,

or $r = 2\cos\theta$.

The point B is on the line $y = x$ which makes an algle $\pi/4$ with OX and so, at B, $\theta = \pi/4$. At the point O of the circle $r = 2\cos\theta$, we have $r = 0$ and so $\theta = \pi/2$. Thus for the region R, r varies from 0 to $2\cos\theta$ and θ varies from $\pi/4$ to $\pi/2$. Also the polar equivalent of dx dy is $r\,d\theta\,dr$.

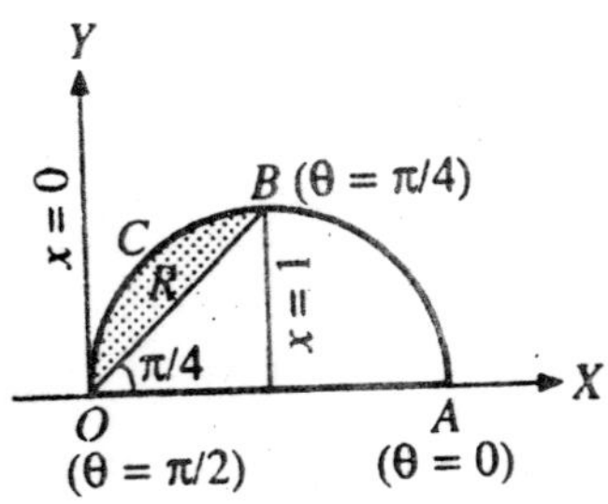

Fig. 1.3

Hence transforming to polar coordinates, we have

$$I = \int_{\theta=\pi/4}^{\pi/2}\int_{r=0}^{2\cos\theta} r^2.r\,d\theta\,de = \int_{\theta=\pi/4}^{\pi/2}\int_{r=0}^{2\cos\theta} r^3 d\theta\,dr$$

$$= \int_{\pi/4}^{\pi/2}\left[\frac{r^4}{4}\right]_{r=0}^{2\cos\theta} d\theta = \frac{16}{4}\int_{\pi/4}^{\pi/2}\cos^4\theta\,d\theta$$

$$= 4\int_{\pi/4}^{\pi/2}\left(\frac{1+\cos 2\theta}{2}\right)^2 d\theta$$

$$= 4\int_{\pi/4}^{\pi/2}(1+2\cos 2\theta+\cos^2 2\theta)d\theta$$

$$= \int_{\pi/4}^{\pi/2}\left[1+2\cos 2\theta+\frac{1+\cos 4\theta}{2}\right]d\theta$$

$$= \int_{\pi/4}^{\pi/2}\left[\frac{3}{2}+2\cos 2\theta+\frac{1}{2}\cos 4\theta\right]d\theta$$

$$= \left[\frac{2}{3}\theta+2.\frac{\sin 2\theta}{2}+\frac{1}{2}.\frac{\sin 4\theta}{4}\right]_{\pi/4}^{\pi/2}$$

$$= \left[\frac{3\pi}{4}-\frac{3\pi}{8}-1\right] = \frac{3\pi}{8}-1.$$

(iii) The given double integral $I = \int_{x=0}^{a}\int_{y=0}^{\sqrt{(a^2+x^2)}} y^2\sqrt{(x^2+y^2)}dx\,dy$.

Here the region of integration R is bounded by $y = 0$, $y = \sqrt{(a^2 - x^2)}$ and $x = 0$, $x = a$. Thus the region of integration R is the area of the circle $x^2 + y^2 = a^2$ lying in the positive quadrant. The polar equation of this circle is $r = a$ and for the region R, r varies from 0 to a and θ varies from 0 to $\pi/2$. Putting $x = r\cos\theta$, $y = r\sin\theta$ and replacing dx dy by $d\theta$ dr, we have

$$I = \int_{\theta=0}^{\pi/2}\int_{r=0}^{a} r^2\sin^2\theta.r.rd\theta\,dr$$

$$= \int_{\theta=0}^{\pi/2}\int_{r=0}^{a} r^4\sin^2\theta d\theta\,dr$$

$$= \int_{0}^{\pi/2}\left[\frac{r^5}{5}\right]_0^a \sin^2\theta\,d\theta = \frac{a^5}{5}\int_0^{\pi/2}\sin^2\theta\,d\theta$$

$$= \frac{a^5}{5}\cdot\frac{1}{2}\cdot\frac{\pi}{2} = \frac{\pi a^5}{20}.$$

Example 11:

Changes the following integrals into polar coordinates.

(i) $\int_0^a \int_0^{\sqrt{(a^2-y^2)}} (x^2+y^2)dy\,dx$.

(ii) $\int_0^{4a} \int_{y^2/4a}^{y} \frac{x^2-y^2}{x^2+y^2} dx\,dy$.

Solution:

(i) Let $I = \int_{y=0}^{a} \int_{x=0}^{\sqrt{(a^2-y^2)}} (x^2+y^2)dx\,dy$.

Changing to polar coordinates, we have

$$I = \int_{\theta=0}^{\pi/2} \int_{r=0}^{a} r^2 . r\,d\theta\,dr$$

$$= \int_{\theta=0}^{\pi/2} \int_{r=0}^{a} r^3\,d\theta\,dr.$$

(ii) Let $I = \int_{y=0}^{4a} \int_{x=y^2/4a}^{y} \frac{x^2-y^2}{x^2+y^2} dx\,dy$.

Here the region of integration R is bound by $x = y^2/4a$, $x = y$, $y = 0$ and $y = 4a$ *i.e.*, the region of integration is the area of the parabola $y^2 = 4ax$ cut off by the straight line $y = x$.

Changing to polar coordinated, the equation $y^2 = 4ax$ becomes

$$(r \sin\theta)^2 = 4a\,(r\cos\theta)$$

or $r = \frac{4a\cos\theta}{\sin^2\theta}$.

At the point B, $\theta = \pi/4$.

At the point O of the parabola

$r = \frac{4a\cos\theta}{\sin^2\theta}$, we have $r = 0$ a nd so $\theta = \pi/2$.

Thus, for the region R, r varies from 0 to $\frac{4a\cos\theta}{\sin^2\theta}$ and θ varies from $\pi/4$ to $\pi/$. Also the polar equivalent of dx dy is $r\,d\theta\,dr$.

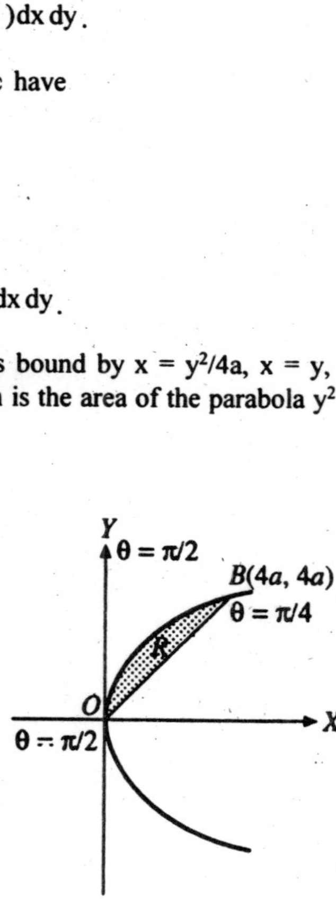

Fig. 1,4

Hence transforming to polar coordinates, we have

$$I = \int_{\theta=\pi/4}^{\pi/2} \int_{r=0}^{(4a\cos\theta)/\sin^2\theta} \frac{r^2(\cos^2\theta - \sin^2)}{r^2} r\,d\theta\,dr$$

$$= \int_{\theta=\pi/4}^{\pi/2} \int_{r=0}^{(4a\cos\theta)/\sin^2\theta} r\cos 2\theta \, d\theta \, dr.$$

Example 12:

Transform to polar coordinates and integrate

$$\iint \sqrt{\left(\frac{1-x^2-y^2}{1+x^2+y^2}\right)} dx\, dy$$

the integral being extended over all positive values of x and y subject to $x^2 + y^2 \le 1$.

Solution:

Here the region of integration R of the given double integral is the area of the circle $x^2 + y^2 = 1$ lying in the positive quadrant. The polar equation of this circle is $r = 1$ and for the region R, r varies from 0 to 1 and θ varies from 0 to $\pi/2$. Also $dx\,dy = r\,d\theta\,dr$.

Hence transforming to polar coordinates, the given double integral

$$= \int_{\theta=0}^{\pi/2} \int_{r=0}^{1} \sqrt{\left(\frac{1-r^2}{1+r^2}\right)} r\, d\theta\, dr$$

$$= \int_{r=0}^{1} r\sqrt{\left(\frac{1-r^2}{1+r^2}\right)} [\theta]_{\theta=0}^{\pi/2} dr,$$

first integrating w.r.t. θ taking r as constant

$$= \frac{\pi}{2}\int_0^1 \frac{r(1-r^2)}{\sqrt{(1-r^4)}} dr$$

$$= \frac{\pi}{2}\int_0^{\pi/2} \frac{(1-\sin t)}{\cos t}\cdot\frac{1}{2}\cos t\, dt,$$

Putting $r^2 = \sin t$ so that $2r\,dr = \cos t\, dt$

$$= \frac{\pi}{4}[t+\cos t]_0^{\pi/2} = \frac{\pi}{4}\left[\frac{\pi}{2}+0-0-1\right] = \frac{\pi}{8}(\pi-2).$$

Example 13:

By changing to polar coordinates, evaluate

$$\iint xy(x^2+y^2)^{n/2} dx\, dy, n+3>0,$$

over the positive quadrant of the circle $x^2 + y^2 = a^2$.

Deduce the value of

$$\iint xy\left(\frac{x^2}{a^2}+\frac{y^2}{b^2}\right)dx\,dy$$

over the positive quadrant of the ellipse $x^2/a^2 + y^2/b^2 = 1$.

Solution:

Let $I = \iint_R xy(x^2+y^2)^{n/2}\,dx\,dy,$

where the region of integration R is the area of the circle $x^2 + y^2 = a^2$ lying in the positive quadrant.

We have $x = r\cos\theta$, $y = r\sin\theta$ and $x^2 + y^2 = a^2$. The polar equation of the circle $x^2 + y^2 = a^2$ is $r = a$ and for the region R, r varies from 0 to a and θ varies from 0 to π/2. Also $dx\,dy = r\,d\theta\,dr$.

∴ transforming to polar coordinates, we have

$$I = \int_{\theta=0}^{\pi/2}\int_{r=0}^{a} r\cos\theta . r\sin\theta . (r^2)^{n/2}\, r\,d\theta\,dr$$

$$= \int_{\theta=0}^{\pi/2}\int_{r=0}^{a} \cos\theta . r^{n+3}\sin\theta\,d\theta\,dr$$

$$= \int_{0}^{\pi/2}\left[\frac{r^n + 4}{n+4}\right]_{r=0}^{a}\cos\theta\sin\theta\,d\theta$$

$$= \frac{a^n + 4}{n+4}\int_0^{\pi/2}\cos\theta\sin\theta\,d\theta$$

$$= \frac{a^n + 4}{n+4}\cdot\frac{1}{2} = \frac{a^n + 4}{2(n+4)}.$$

Now let $I_1 = \iint xy\left(\frac{x^2}{a^2}+\frac{y^2}{b^2}\right)dx\,dy$, the integral being extended over the positive quadrant of the ellipse $\frac{x^2}{a^2}+\frac{y^2}{b^2} = 1$.

Put $\frac{x}{a} = u, \frac{y}{b} = v$. Then $dx = a\,du$, $dy = b\,dv$.

∴ $I_1 = \iint au.bv.(u^2+v^2)\,ab\,du\,dv$, the integral being extended to all positive values of u and v subject to the condition $u^2 + v^2 \leq 1$

$$= a^2b^2\iint uv(u^2+v^2)\,du\,dv.$$

Now putting $a = 1$ and $n = 2$ in the value of I, we have

$$\iint uv(u^2+v^2)\,du\,dv = \frac{1}{12}.$$

$$\therefore I_1 = a^2b^2.\frac{1}{12} = \frac{a^2b^2}{12}.$$

Example 14:

Evaluate $\int_{y=0}^{3}\int_{x=0}^{2}\int_{z=0}^{1}(x+y+z)\,dz\,dx\,dy$.

Solution:

The given integral

$$= \int_{y=0}^{3}\int_{x=0}^{2}\left\{\int_{0}^{1}(x+y+z)\,dz\right\}dx\,dy$$

$$= \int_{y=0}^{3}\int_{x=0}^{2}\left\{xz+yz+\frac{z^2}{2}\right\}_0^1 dx\,dy = \int_0^3\left\{\int_0^2\left(x+y+\frac{1}{2}\right)dx\right\}dy$$

$$= \int_0^3\left\{\frac{x^2}{2}+xy+\frac{x}{3}\right\}_0^2 dy = \int_0^3(3+2y)\,dy = \left[3y+\frac{2y^2}{2}\right]_0^3 = 18.$$

Example 15:

Show that $\int_{x=0}^{1}\int_{y=0}^{2}\int_{z=1}^{2}x^2yz\,dz\,dy\,dx = 1$.

Solution:

We have $\int_{x=0}^{1}\int_{y=0}^{2}\int_{z=1}^{2}x^2yz\,dz\,dy\,dx$

$$= \int_{x=0}^{1}\int_{y=0}^{2}\left\{\int_{1}^{2}x^2yz\,dz\right\}dy\,dx$$

$$= \int_{x=0}^{1}\int_{y=0}^{2}\left[x^2y.\frac{z^2}{2}\right]_1^2 dy\,dx = \frac{1}{2}\int_0^1\left[\int_0^2(3x^2y)\,dy\right]dx$$

$$= \frac{3}{2}\int_0^1\left[x^2\cdot\frac{y^2}{2}\right]_0^2 dx = \frac{3}{4}\int_0^1 4x^2dx = 3\left[\frac{x^3}{3}\right]_0^1 = 3.\frac{1}{3} = 1.$$

Example 16:

Evaluate $\int_0^1\int_0^1\int_0^1 e^{x+y+z}\,dx\,dy\,dz$.

Solution:

$$\int_0^1\int_0^1\int_0^1 e^{x+y+z}\,dx\,dy\,dz = \int_0^1\int_0^1\left[\int_0^1 e^{x+y+z}\,dx\right]dy\,dz$$

$$= \int_0^1\int_0^1\left[\int_0^1 e^{x+y+z}\right]_0^1 dy\,dz = \int_0^1\left[\int_0^1\{e^{1+y+z} - e^{y+z}\}dy\right]dz$$

$$= \int_0^1\left[e^{1+y+z} - e^{y+z}\right]_0^1 dz$$

$$= \int_0^1\{(e^{2+z} - e^{1+z}) - (e^{1+z} - e^z)\}dz$$

$$= \int_0^1(e^{2+z} - 2e^{1+z} + e^z)dz = \int_0^1(e^2 - 2e + 1)e^z dz$$

$$= (e^2 - 2e + 1)\int_0^1 e^z dz = (e-1)^2\left[e^z\right]_0^1 = (e - 1)^2 (e - e^0)$$

$$= (e - 1)^2 (e - 1) = (e - 1)^3.$$

Example 17:

Evaluate $\int_{-1}^1\int_0^z\int_{x-z}^{x+z}(x + y + z)dy\,dx\,dz.$

Solution:

Here $x - z$ to $x + z$ are the limits of integration of y, 0 to z are those of x and -1 to 1 are those of z. The given triple integral is

$$= \int_{-1}^1\int_0^z\left[\int_{x-z}^{x+z}(x + y + z)dy\right]dx\,dz$$

$$= \int_{-1}^1\int_0^z\left[xy + \frac{y^2}{2} + zy\right]_{x-z}^{x+z} dx\,dz$$

$$= \int_{-1}^1\int_0^z\left[x(x + z) + \frac{(x + z)^2}{2} + z(x + z) - x(x - z)\right.$$

$$\left.- \frac{(x - z)^2}{2} - z(x - z)\right]dx\,dz$$

$$= \int_{-1}^1\left[\int_0^z(4xz + 2z^2)dx\right]dz = \int_{-1}^1\left[2zx^2 + 2z^2x\right]_0^z dz$$

$$= \int_{-1}^1(2z.z^2 + 2z^2.z)dz = 4\int_{-1}^1 z^3 dz$$

$$= 4\left[\frac{z^4}{4}\right]_{-1}^1 = 1.\ [1 - 1] = 0.$$

Example 18:

Evaluate the following integrals.

(i) $\int_0^1 \int_0^{1-x} \int_0^{1-x-y} x\,yz\,dx\,dy\,dz;$

(ii) $\int_{-c}^{c} \int_{-b}^{b} \int_{-a}^{a} (x^2 + y^2 + z^2)dx\,dy\,dz;$

(iii) $\int_0^{\log 2} \int_0^{x} \int_0^{x+\log y} . e^{x+y+z} dx\,dy\,dz;$

(iv) $\int_0^1 \int_{y^2}^{1} \int_0^{1-x} x\,dy\,dx\,dz.$

(v) $\int_0^1 \int_0^{1-x} \int_0^{1-x-y} \frac{dx\,dy\,dz}{(1+x+y+z)^3}.$

(vi) $\int_0^1 \int_0^{\sqrt{(1-x^2)}} \int_0^{\sqrt{(1-x^2-y^2)}} xyz\,dz\,dy\,dx$

(vii) $\int_1^e \int_1^{\log y} \int_1^{e^x} \log z\,dz\,dx\,dy.$

Solution:

(i) We have $\int_0^1 \int_0^{1-x} \int_0^{1-x-y} x\,yz\,dx\,dy\,dz$

$$= \int_0^1 \int_0^{1-x} xy \left[\frac{x^2}{2} \right]_0^{1-x-y} dx\,dy,$$

integrating w.r.t. z regarding x and y as constants

$$= \frac{1}{2} \int_0^1 \int_0^{1-x} xy \{(1-x) - y\}^2 dx\,dy$$

$$= \frac{1}{2} \int_0^1 \int_0^{1-x} x[y(1-x)^2 - 2(1-x)y^2 + y^3]dx\,dy$$

$$= \frac{1}{2} \int_0^1 x \left[\frac{(1-x)^2 y^2}{2} - \frac{2(1-x)y^3}{3} + \frac{y^4}{4} \right]_0^{1-x} dx$$

integrating w.r.t. y regarding x as constants

$$= \frac{1}{24} \int_0^1 x[6(1-x)^4 - 8(1-8)^4 + 3(1-x)^4]\,dx$$

$$= \frac{1}{24} \int_0^1 x(1-x)^4 dx$$

$$= \frac{1}{24} \int_0^{\pi/2} \sin^2\theta \cos^8\theta . 2\sin\theta\cos\theta\,d\theta,$$

putting $x = \sin^2\theta$ so that $dx = 2\sin\theta\cos\theta\,d\theta$

$$= \frac{1}{12}\int_0^{\pi/2} \sin^3\theta\cos^9\theta\, d\theta = \frac{1}{12}\cdot\frac{2.8.6.4.2}{12.10.8.6.4.2} = \frac{1}{720}.$$

(ii) Here the integrand $x^2 + y^2 + z^2$ is a symmetrical expression in x, y and z and therefore the limits of integration can be as signed at pleasure. We have the given integral

$$= \int_{z=-c}^{c}\int_{y=-b}^{b}\int_{x=-a}^{a}(x^2+y^2+z^2)dx\,dy\,dz$$

$$= 2\int_{z=-c}^{c}\int_{y=-b}^{b}\int_{x=0}^{a}(x^2+y^2+z^2)dx\,dy\,dz,$$

because $x^2 + y^2 + z^2$ is an even function of x

$$= 2\int_{z=-c}^{c}\int_{y=-b}^{b}\left[\frac{x^3}{3}+(y^2+z^2)x\right]_0^a dy\,dz.$$

integrating w.r.t. x regarding y and z as constants

$$= 2\int_{z=-c}^{c}\int_{y=-b}^{b}\left[\frac{a^3}{3}+ay^2+az^2\right]dy\,dz$$

$$= 4\int_{z=-c}^{c}\int_{0}^{b}\left[\frac{a^3}{3}+az^2+ay^2\right]dy\,dz$$

because $\frac{a^3}{3}+az^2+ay^2$ is an even function of y

$$= 4\int_{z=-c}^{c}\left[\frac{a^3}{3}y+az^2y+\frac{ay^3}{3}\right]dz,$$

integrating w.r.t. y regarding z as constant

$$= 4\int_{z=-c}^{c}\left[\frac{a^3b}{3}+abz^2+\frac{ab^3}{3}\right]dz = 8\int_0^c\left[\frac{a^3b}{3}+abz^2+\frac{ab^3}{3}\right]dz$$

$$= 8\left[\frac{a^3b}{3}z+ab\frac{z^3}{3}+\frac{ab^3}{3}z\right]_0^c$$

$$= \frac{8}{3}(a^2bc+abc^3+ab^3c) = \frac{8}{3}abc(a^2+b^2+c^2).$$

(iii) We have $\int_0^{\log 2}\int_0^{x}\int_0^{x+\log y} e^{x+y+z}dx\,dy\,dz$

$$= \int_0^{\log 2}\int_0^{x}\left[e^{x+y+z}\right]_0^{x+\log y} dx\,dy,$$

integrating w.r.t. z regarding x and y as constants

$$= \int_0^{\log 2}\int_0^x [e^{x+y+x+\log y} - e^{x+y}]dx\,dy$$

$$= \int_0^{\log 2}\int_0^x [e^{2x}e^y e^{\log y} - e^x e^y]dx\,dy$$

$$= \int_0^{\log 2}\int_0^x [e^{2x}ye^y - e^x e^y]dx\,dy,$$

$$= \int_0^{\log 2}\left[\int_0^x e^{2x}y\,e^y\,dy - \int_0^x e^x e^y\,dy\right]dx$$

$$= \int_0^{\log 2}\left[e^{2x}\left\{ye^y\right\}_0^x - e^{2x}\int_0^x e^y\,dy - e^x\left\{e^y\right\}_0^x\right]dx$$

integrating w.r.t. y regarding x as a constant; to integrate ye^y we have applied integration by parts

$$= \int_0^{\log 2}\left[e^{2x}.x\,e^x - e^{2x}\left\{e^y\right\}_0^x - e^x(e^x - 1)\right]dx$$

$$= \int_0^{\log 2}[xe^{3x} - e^{2x}(e^x - 1) - e^{2x} + e^x]dx$$

$$= \int_0^{\log 2}[x\,e^{3x} - e^{3x} + e^x]dx$$

$$= \int_0^{\log 2}x\,e^{3x}dx - \int_0^{\log 2}e^{3x}dx + \int_0^{\log 2}e^x dx$$

$$= \frac{1}{3}\left[x\,e^{3x}\right]_0^{\log 2} - \frac{1}{3}\int_0^{\log 2}e^{3x}dx - \int_0^{\log 2}e^{3x}dx + \int_0^{\log 2}e^x dx$$

$$= \frac{1}{3}(\log 2)e^{3\log 2} - \frac{4}{3}\left[\frac{e^{3x}}{3}\right]_0^{\log 2} + \left[e^x\right]_0^{\log 2}$$

$$= \frac{1}{3}(\log 2)e^{3\log 8} - \frac{4}{9}(e^{3\log 2} - 1) + (e^{\log 2} - 1)$$

$$= \frac{8}{3}\log 2 - \frac{4}{9}(8-1) + (2-1) = \frac{8}{3}\log 2 - \frac{28}{9} + 1$$

$$= \frac{8}{3}\log 2 - \frac{19}{9}.$$

(iv) We have $\int_0^1\int_{y^2}^1\int_0^{1-x} x\,dy\,dx\,dz$

$$= \int_0^1\int_{y^2}^1 x[z]_0^{1-x}\,dy\,dx,$$

integrating w.r.t. z regarding x and y as constants

$$= \int_0^1 \int_{y^2}^1 x(1-x)\,dy\,dx$$

$$= \int_0^1 \left[\frac{x^2}{2} + \frac{x^3}{3}\right]_{y^2}^1 dy,$$

integrating w.r.t. x regarding y as constant

$$= \int_0^1 \left[\frac{1}{2} - \frac{1}{3} - \frac{y^4}{2} + \frac{y^6}{3}\right] dy = \int_0^1 \left[\frac{1}{6} - \frac{y^4}{2} + \frac{y^6}{3}\right] dy$$

$$= \left[\frac{1}{6}y - \frac{y^5}{10} + \frac{y^7}{21}\right]_0^1 = \frac{1}{6} - \frac{1}{10} + \frac{1}{21} = \frac{35-21+10}{210} = \frac{24}{210} = \frac{4}{35}.$$

(v) We have $\int_0^1 \int_0^{1-x} \int_0^{1-x-y} \frac{dx\,dy\,dz}{(1+x+y+z)^3}$

$$= \int_0^1 \int_0^{1-x} \left[-\frac{1}{2(1+x+y+z)^2}\right]_0^{1-x-y} dx\,dy$$

$$= \frac{1}{2}\int_0^1 \int_0^{1-x} \left[-\frac{1}{4} + \frac{1}{(1+x+y+z)^2}\right] dx\,dy$$

$$= \frac{1}{2}\int_0^1 \left[-\frac{1}{4}y - \frac{1}{(1+x+y)}\right]_0^{1-x} dx$$

$$= \frac{1}{2}\int_0^1 \left[-\frac{1}{4}(1-x) - \frac{1}{2} + \frac{1}{(1+x)}\right] dx$$

$$= \frac{1}{2}\int_0^1 \left[-\frac{3}{4} + \frac{1}{4}x + \frac{1}{(1+x)}\right] dx$$

$$= \frac{1}{2}\left[-\frac{3}{4}x + \frac{1}{4}\cdot\frac{x^2}{2} + \log(1+x)\right]_0^1$$

$$= \frac{1}{2}\left[-\frac{3}{4} + \frac{1}{8} + \log 2\right] = \frac{1}{2}\left(\log 2 - \frac{5}{8}\right).$$

(vi) The given integral I

$$= \int_{x=0}^1 \int_{y=0}^{\sqrt{(1-z^2)}} \int_{z=0}^{\sqrt{(1-x^2-y^2)}} xyz\,dz\,dy\,dx$$

$$= \int_{x=0}^1 \int_{y=0}^{\sqrt{(1-z^2)}} xy\left[\frac{z^2}{2}\right]_{z=0}^{\sqrt{(1-x^2-y^2)}} dy\,dx$$

$$= \int_{x=0}^{1}\int_{y=0}^{\sqrt{(1-z^2)}} \frac{1}{2} xy(1-x^2-y^2)\,dy\,dx$$

$$= \int_{x=0}^{1} \frac{1}{2} x\left[(1-x^2)\frac{y^2}{2} - \frac{y^4}{4}\right]_{y=0}^{\sqrt{(1-x^2)}} dx$$

$$= \int_0^1 \frac{1}{2} x\left[\frac{1}{2}(1-x^2)^2 - \frac{1}{4}(1-x^2)^2\right] dx$$

$$= \int_0^1 \frac{1}{2} x \cdot \frac{1}{4}(1-x^2)^2\,dx$$

$$= \frac{1}{8}\int_0^{\pi/2} \sin\theta.\cos^4\theta\cos\theta\,d\theta,$$

putting $x = \sin\theta$ so that $dx = \cos\theta\,d\theta$

$$= \frac{1}{8}\int_0^{\pi/2} \sin\theta\cos^5\theta\,d\theta = \frac{1}{8}\cdot\frac{1.4.2}{6.4.2} = \frac{1}{48}.$$

(vii) The given integral I

$$= \int_{y=1}^{e}\int_{x=1}^{\log y}\int_{z=1}^{e^x} \log z\,dz\,dx\,dy$$

$$= \int_{y=1}^{e}\int_{x=1}^{\log y} [z\log z - z]_{z=1}^{e^x}\,dx\,dy$$

$$= \int_{y=1}^{e}\int_{x=1}^{\log y} [xe^x - e^x + 1]dx\,dy$$

$$= \int_{y=1}^{e} [xe^x - 2e^x + x]_{x=1}^{\log y}\,dy$$

$$= \int_1^e [y\log y - 2y + \log y - e + 2e - 1]dy \qquad [\because e^{\log y} = y]$$

$$= \int_1^e [y\log y + \log y - 2y + e - 1]dy$$

$$= \left[\frac{y^2}{2}\log y - \frac{y^2}{4} + y\log y - y - y^2 + ey - y\right]_1^e$$

$$= \frac{1}{2}e^2 - \frac{1}{4}e^2 + e - e - e^2 + e^2 - e + \frac{1}{4} + 1 + 1 - e + 1s$$

$$= \frac{1}{4}e^2 - 2e + \frac{13}{4}.$$

Example 19:

Evaluate $\int_1^3\int_{1/x}^1\int_0^{\sqrt{(xy)}} xyz\,dx\,dy\,dz$.

Solution:

The given triple integral is

$$\int_1^3\int_{1/x}^1\left[\int_0^{\sqrt{(xy)}} xyz\,dz\right]dx\,dy = \int_1^3\int_{1/x}^1\left[xy.\frac{z^2}{2}\right]_0^{\sqrt{(xy)}} dx\,dy$$

$$= \frac{1}{2}\int_1^3\left[\int_{1/x}^1 x^2y^2dy\right]dx = \frac{1}{2}\int_1^3\left[x^2.\frac{y^3}{3}\right]_{1/x}^1$$

$$= \frac{1}{6}\int_1^3\left[x^2 - \frac{1}{x}\right]dx = \frac{1}{6}\left[\frac{x^3}{3} - \log x\right]_1^3$$

$$= \frac{1}{6}\left[(9 - \log 3) - \left(\frac{1}{3} - \log 1\right)\right] = \frac{1}{6}\left[\left(9 - \frac{1}{3}\right) - \log 3\right]$$

$$= \frac{1}{6}\left[\frac{26}{3} - \log 3\right].$$

Example 20:

Evaluate $\int_0^{\pi/2} d\theta\int_0^{a\sin\theta} dr\int_0^{(a^2-r^2)/a} r\,dz.$

Solution:

The given triple integral is

$$= \int_0^{\pi/2} d\theta\int_0^{a\sin\theta} dr[rz]_0^{(a^2-r^2)/a}$$

$$= \int_0^{\pi/2} d\theta\int_0^{a\sin\theta} \frac{r(a^2 - r^2)}{a}dr$$

$$= \frac{1}{a}\int_0^{\pi/2}\left[\frac{a^2r^2}{2} - \frac{r^2}{4}\right]_0^{a\sin\theta} d\theta = \frac{a^3}{4}\int_0^{\pi/2}(2\sin^2\theta - \sin^4\theta)d\theta$$

$$= \frac{a^3}{4}\left[2\cdot\frac{1}{2}\cdot\frac{\pi}{2} - \frac{3}{4}\cdot\frac{1}{2}\cdot\frac{\pi}{2}\right] = \frac{5a^3\pi}{64}.$$

Example 21:

Evaluate $\int_0^a\int_0^x\int_0^{x+y} e^{x+y+z}dx\,dy\,dx$.

Solution:

The given triple integral is

$$= \int_0^a \int_0^x \left[\int_0^{x+y} e^{x+y+z} dz \right] dx\, dy = \int_0^a \int_0^x \left[e^{x+y+z} \right]_{z=0}^{x+y} dx\, dy$$

$$= \int_0^a \int_0^x [e^{2(x+y)} - e^{(e+y)}] dx\, dy = \int_0^a \left[\frac{1}{2} e^{2(x+y)} - e^{(x+y)} \right]_0^x dx$$

$$= \int_0^a \left[\frac{1}{2}(e^{4x} - e^{2x}) - (e^{2x} - e^x) \right] dx = \int_0^a \left(\frac{1}{2} e^{4x} - \frac{3}{2} e^{2x} + e^x \right) dx$$

$$= \left[\frac{1}{2} \cdot \frac{1}{4} e^{4x} - \frac{3}{4} \cdot \frac{1}{2} e^{2x} + e^x \right]_0^a$$

$$= \left[\left(\frac{1}{8} e^{4a} - \frac{3}{4} e^{2a} + e^a \right) - \left(\frac{1}{8} e^0 - \frac{3}{4} e^0 + e^0 \right) \right]$$

$$= \frac{1}{8} e^{4a} - \frac{3}{4} e^{2a} + e^a - \left(\frac{1}{8} - \frac{3}{4} + 1 \right) = \frac{1}{8}(e^{4a} - 6e^{2a} + 8e^a - 3)$$

Example 22:

Evaluate $\int_0^4 \int_0^{2\sqrt{z}} \int_0^{\sqrt{(4z-x^2)}} dz\, dx\, dy$.

Solution:

The given triple integral is

$$= \int_0^4 \int_0^{2\sqrt{z}} \left[\int_0^{\sqrt{(4z-x^2)}} dy \right] dz\, dx = \int_0^4 \int_0^{2\sqrt{z}} [y]_0^{\sqrt{(4z-x^2)}} dz\, dx$$

$$= \int_0^4 \left[\int_0^{2\sqrt{z}} \sqrt{(4z - x^2)} dx \right] dz$$

$$= \int_0^4 \left[\frac{x}{2} \sqrt{(4z - x^2)} + \frac{4z}{2} \sin^{-1} \frac{x}{2\sqrt{z}} \right]_0^{2\sqrt{z}} dz$$

$$= \int_0^4 \left[0 + \frac{4z}{2} \sin^{-1} \frac{2\sqrt{z}}{2\sqrt{z}} \right]_0^{2\sqrt{z}} dz = \int_0^4 2z . \frac{\pi}{2} dz = \int_0^4 \pi z\, dz$$

$$= \pi \left[\frac{z^2}{2} \right]_0^4 = \frac{\pi}{2}[16] = 8\pi .$$

Example 23:

Evaluate $\int_0^a \int_0^{a-x} \int_0^{a-x-y} x^2 dx\, dy\, dz$.

Solution:

The given triple integral

$$= \int_0^a \int_0^{a-x} \int_0^{a-x-y} x^2 dx\, dy\, dz = \int_0^a \int_0^{a-x} x^2 [z]_0^{a-x-y} dx\, dy$$

integrating w.r.t. z regarding x and y as constants

$$= \int_0^a \int_0^{a-x} x^2 [a-x-y] dx\, dy = \int_0^a \int_0^{a-x} x^2 [(a-x)-y] dx\, dy$$

$$= \int_0^a \int_0^{a-x} x^2 \left[(a-x)y - \frac{1}{2}y^2 \right]_0^{a-x} dx,$$

integrating w.r.t. y regarding x as constant

$$= \int_0^a x^2 [(a-x)^2 - \frac{1}{2}(a-x)^2 dx$$

$$= \int_0^a x^2 . \frac{1}{2}(a-x)^2 dx = \frac{1}{2}\int_0^a x^2 (a^2 - 2ax + x^2) dx$$

$$= \frac{1}{2}\int_0^a (x^2a^2 - 2a x^3 + x^4) dx = \frac{1}{2}\left[a^2 \frac{x^3}{3} - 2a\frac{x^4}{4} + \frac{x^5}{5} \right]_0^a$$

$$= \frac{1}{2}\left[\frac{1}{3}a^5 - \frac{1}{2}a^5 + \frac{1}{5}a^5 \right] = \frac{1}{2}\left(\frac{1}{3} - \frac{1}{2} + \frac{1}{5} \right) a^5 = \frac{1}{60}a^5.$$

Example 24:

Evaluate $\iint (x+y+a) dx\, dy$ *over the circular area* $x^2 + y^2 \le a^2$.

Solution:

Here the region of integration R can be expressed as $-a \le y \le a$, $-\sqrt{(a^2 - y^2)} \le x \le \sqrt{(a^2 - y^2)}$,

where the first integration is to be performed w.r.t. x regarding y as constant.

$$\therefore \iint_R (x+y+a) dx\, dy$$

$$= \int_{y=-a}^{a} \int_{x=-\sqrt{(a^2-y^2)}}^{\sqrt{(a^2-y^2)}} (x+y+a) dx\, dy$$

$$= \int_{-a}^{a} \left[\frac{x^2}{2} + (y+a)x \right]_{-\sqrt{(a^2-y^2)}}^{\sqrt{(a^2-y^2)}} dy,$$

[integrating w.r.t. x treating y as a constant]

$$= \int_{-a}^{a}\left[\frac{a^2-y^2}{2}+(y+a)\sqrt{(a^2-y^2)}dy - \frac{a^2-y^2}{2}+(y+a)\sqrt{(a^2-y^2)}\right]$$

$$= \int_{-a}^{a} 2(y+a)\sqrt{(a^2-y^2)}dy$$

$$= \int_{-a}^{a} 2y.\sqrt{(a^2-y^2)}dy + 2a\int_{-a}^{a}\sqrt{(a^2-y^2)}dy$$

$$= 0 + 2a.\ 2\int_0^a\sqrt{(a^2-y^2)}dy,$$

the first integal vanishes because the integrand is an odd function of y

$$= 4a\left[\frac{y\sqrt{(a^2-y^2)}}{2}+\frac{a^2}{2}\sin^{-1}\frac{y}{a}\right]_0^a$$

$$= 4a\left[0+\frac{1}{2}a^2\sin^{-1}1-0\right]$$

$$= 4a.\ \frac{1}{2}a^2.\frac{1}{2}\pi = \pi a^3.$$

Example 25:

Evaluate $\iint x^2y^2\,dxdy$ *over the region bounded by* $x = 0$, $y = 0$ *and* $x^2 + y^2 = 1$.

Solution:

The given region for integration is the area of the positive quadrant of the circel $x^2 + y^2 = 1$ in the xy-plane. This region R can be expressed either as

$$0 \le x \le \sqrt{(1-y^2)},\ 0 \le y \le 1$$

or as $\quad 0 \le y \le \sqrt{(1-x^2)},\ 0 \le x \le 1.$

the first integration to be performed w.r.t. x regarding y as constant

$$= \int_0^1 y^2\left[\frac{x^3}{3}\right]_0^{\sqrt{(1-y^2)}} dy = \int_0^1 \frac{1}{2}y^2(1-y^2)^{3/2}\,dy.$$

Put $\quad y = \sin\theta$ so that $dy = \cos\theta\, d\theta$.

When $y = 0$, $\theta = 0$ and when $y = 1$, $\theta = \pi/2$.

$$\therefore \iint_R x^2y^2dx\,dy = \int_0^{\pi/2}\frac{1}{3}\sin^2\theta(1-\sin^2\theta)^{3/2}.\cos\theta\, d\theta$$

$$= \frac{1}{3}\int_0^{\pi/2} \sin^2\theta \cos^4\theta \, d\theta = \frac{1}{3}\cdot\frac{1.3.1}{6.4.2}\cdot\frac{\pi}{2} = \frac{\pi}{96}.$$

Example 26:

Evaluate $\iint x^2y^2 \, dx\, dy$ *over the region* $x^2 + y^2 \le 1$.

Solution:

Here the given region of integration R is the whole area of the circle $x^2 + y^2 = 1$. This region R can be expressed as $-\sqrt{(1-y^2)} \le x \le \sqrt{(1-y^2)}$, $-1 \le y \le 1$.

$$\therefore \iint_R x^2y^2 \, dx\, dy = \int_{y=-1}^{1}\int_{x=-\sqrt{(1-y)}}^{\sqrt{(1-y^2)}} x^2y^2 \, dx\, dy$$

$$= \int_{y=-1}^{1} \int_{x=0}^{\sqrt{(1-y^2)}} x^2y^2 \, dxdy,$$

by a property of definite integrals because x^2 is an even function of x

$$= 2\int_{-1}^{1} \frac{1}{3} y^2(1-y^2)^{3/2} dy, \text{ proceeding as in Ex. 12 (a)}$$

$$= \frac{4}{3}\int_0^1 y^2(1-y^2)^{3/2} dy$$

because $y^2 (1 - y^2)^{3/2}$ is an even function of y

$$= \frac{\pi}{24}.$$

Example 27:

Find the area of the ellipse $x^2/a^2 + y^2/b^2 = 1$, *by double integration.*

Solution:

From the equation of the ellipse, we have

$$\frac{y}{b} = \pm\sqrt{\left\{1 - \frac{x^2}{a^2}\right\}}.$$

So the region of integration R to cover the area of the ellipse can be considered as bounded by $y = -b\sqrt{(1 - x^2/a^2)}$, $y = b\sqrt{(1 - x^2/a^2)}$, $x = -a$ and $x = a$.

Therefore the required area of the ellipse

$$= \iint_R dx\, dy = \int_{x=-a}^{a}\int_{y=-b\sqrt{(1-x^2/a^2)}}^{b\sqrt{(1-x^2/a^2)}} 1.dx\, dy$$

$$= \int_{-a}^{a}\left[2\int_{0}^{b\sqrt{(1-x^2/a^2)}} 1.dy\right]dy = 2\int_{-a}^{a}[y]_{0}^{b\sqrt{(1-x^2/a^2)}}dx$$

$$= 2\int_{-a}^{a} b\sqrt{\left(1-\frac{x^2}{a^2}\right)}dx = 2.2\int_{0}^{a} b\sqrt{\left(1-\frac{x^2}{a^2}\right)}dx$$

$$= \frac{4b}{a}\int_{0}^{a}\sqrt{(a^2-x^2)}dx = \frac{4b}{a}\left[\frac{x\sqrt{(a^2-x^2)}}{2}+\frac{a^2}{2}\sin^{-1}\frac{x}{a}\right]_{0}^{a}$$

$$= \frac{4b}{a}\left[0+\frac{a^2}{2}\{\sin^{-1}1-\sin^{-1}0\}\right] = \frac{4b}{a}\cdot\frac{a^2}{2}\cdot\frac{\pi}{b} = \pi ab.$$

Example 28:

Compute the value of $\iint_R y\,dx\,dy$, *where R its the region in the first quadrant bounded by the ellipse* $x^2/a^2 + y^2/b^2 = 1$.

Solution:

If the first integration is to be performed w.r.t. y regarding x as a constant, then the given region of integration can be expressed as $0 \le x \le a$, $0 \le y \le b\sqrt{(1 - x^2/a^2)}$.

$$\therefore \iint_R y\,dx\,dy = \int_{x=0}^{a}\int_{y=0}^{b\sqrt{(1-x^2/a^2)}} y\,dx\,dy$$

$$= \int_{0}^{a}\left[\frac{y^2}{2}\right]_{0}^{b\sqrt{(1-x^2/a^2)}}dx = \frac{1}{2}\int_{0}^{a} b^2\left(1-\frac{x^2}{a^2}\right)dx$$

$$= \frac{b^2}{2a^2}\int_{0}^{a}(a^2-x^2)dx = \frac{b^2}{2a^2}\left[a^2x-\frac{x^3}{3}\right] = \frac{b^2}{2a^2}\cdot\frac{2a^3}{3} = \frac{ab^2}{3}.$$

Example 29:

Evaluate $\iint (x+y)^2$ *dx dy over the area bounded by the ellipse* $x^2/a^2 + y^2/b^2 = 1$. *Hence find the mass of an elliptic plate whose density per unit area is given by* $\rho = k\,(x+y)^2$.

Solution:

The region of integration can be considered as bounded by $y = -b\sqrt{(1\ x^2/a^2)}$, $y = b\sqrt{(1-x^2/a^2)}$, $y = b\sqrt{(1-x^2/a^2)}$, $x = -a$ and $x = a$.

$$\therefore \iint (x+y)^2\,dx\,dy = \int_{-a}^{a}\int_{-b\sqrt{(1-x^2/a^2)}}^{b\sqrt{(1-x^2/a^2)}}(x^2+y^2+2xy)dx\,dy,$$

the first integration to be performed w.r.t. y regarding x as a constant

$$= \int_{-a}^{a} 2\int_{0}^{b\sqrt{(1-x^2/a^2)}} (x^2 + y^2)dx\,dy,$$

[$\therefore$ 2xy being an odd function of y, its integration under the giver limits of y is 0]

$$= 2\int_{-a}^{a}\left[x^2y + \frac{y^2}{3}\right]_0^{b\sqrt{(1-x^2/a^2)}} dx$$

$$= 2\int_{-a}^{a}\left\{x^2 b\sqrt{\left(1-\frac{x^2}{a^2}\right)} + \frac{b^3}{3}\left(1-\frac{x^2}{a^2}\right)^{3/2}\right\}dx$$

$$= 4\int_{0}^{a}\left\{x^2 b\sqrt{\left(1-\frac{x^2}{a^2}\right)} + \frac{b^3}{3}\left(1-\frac{x^2}{a^2}\right)^{3/2}\right\}dx$$

$$= 4b\int_0^{\pi/2}\left\{a^2\sin^2\theta\cos\theta + \frac{b^2}{3}\cos^3\theta\right\}a\cos\theta\,d\theta,$$

putting $x = a\sin\theta$ so that $dx = a\cos\theta\, d\theta$

$$= 4ab\int_0^{\pi/2}\left\{a^2\sin^2\theta\cos^2\theta + \frac{b^2}{3}\cos^4\theta\right\}d\theta$$

$$= 4ab\left[a^2\int_0^{\pi/2}\sin^2\theta\cos^2\theta\,d\theta + \frac{b^2}{3}\int_0^{\pi/2}\cos^4\theta\,d\theta\right]$$

$$= 4ab\left[a^2.\frac{1.1}{4.2}.\frac{\pi}{2} + \frac{b^2}{3}.\frac{3.1}{4.2}.\frac{\pi}{2}\right], \text{[by Walli's formula]}$$

$$= 4ab\left[\frac{1}{16}\pi a^2 + \frac{1}{16}\pi b^2\right] = \frac{1}{4}\pi\, ab(a^2 + b^2).$$

The mass of an elliptic plate whose density is given by

$\rho = k\ (x + y)^2$

$= \iint_A k(x+y)^2\,dx\,dy$, where the integrations to be performed over the area A of the ellipse

$$= k.\frac{1}{4}.\pi\, ab(a^2 + b^2).$$

Example 30:

Evaluate $\iint xy\,dx\,dy$ over the region in the positive quadrant for which $x + y \le 1$.

solution:

The region of integration is the area bounded by the lines x = 0, y = 0 and x + y = 1.

To cover this region of integration R, x varies from 0 to 1 and y varies from 0 to 1 − x.

$$\therefore \iint_R xy\,dx\,dy = \int_{x=0}^{1}\int_{y=0}^{1-x} xy\,dx\,dy = \int_0^1 x\left[\frac{y^2}{2}\right]_0^{1-x} dx$$

$$=\frac{1}{2}\int_0^1 x(1-x)^2\,dx = \frac{1}{2}\int_0^1 x(1-2x+x^2)\,dx = \frac{1}{2}\left[\frac{x^2}{2}-2.\frac{x^3}{3}+\frac{x^4}{4}\right]_0^1$$

$$=\frac{1}{2}\left[\frac{1}{2}-\frac{2}{3}+\frac{1}{4}\right] = \frac{1}{24}.$$

Example 31:

Evaluate $\iint e^{2x+3y}\,dx\,dy$ *over the triangle bounded by* $x = 0$, $y = 0$ *and* $x + y = 1$.

Solution:

The given region of integration R can be expressed as $0 \le x \le 1$, $0 \le y \le 1 - x$,

where the first integration is to be performed w.r.t. y regarding x as a constant.

$$\therefore \iint_R e^{x+3y}\,dx\,dy = \int_0^1\int_0^{1-x} e^{2x+3y}\,dx\,dy$$

$$=\int_0^1\left[\frac{e^{2x+3y}}{3}\right]_0^{1-x} dx = \frac{1}{3}\int_0^1 [e^{3-x}-e^{2x}]\,dx$$

$$=\frac{1}{3}\left[-e^{3-x}-\frac{e^{2x}}{3}\right]_0^1 = -\frac{1}{3}[(e^2-e^3)+\frac{1}{2}(e^2-e^0)]$$

$$=-\frac{1}{3}[-e^2(e-1)+\frac{1}{2}(e+1)(e-1)] = \frac{1}{3}(e-1)[e^2-\frac{1}{2}(e+1)]$$

$$=\frac{1}{6}(e-1)(2e^2-e-1) = \frac{1}{6}(e-1)\{(e-1)(2e+1)\}$$

$$=\frac{1}{6}(e-1)^2(2e+1).$$

Example 32:

Evaluate $\iint (x^2 + y^2)\,dx\,dy$ *over the region in the positive quadrant for which* $x + y \le 1$.

Solution:

The region of integration R is the area bounded by the coordinate axes and the straight line x + y = 1. Therefore, the region R is bounded by y = 0, y = 1 = x and x = 0, x = 1.

Therefore $\iint_R (x^2 + y^2)\,dx\,dy = \int_{x=0}^{1}\int_{y=0}^{1-x} (x^2 + y^2)\,dx\,dy$,

the first integration to be performed w.r.t. y regarding x as constant

$$= \int_0^1 \left[x^2 y + \frac{y^2}{3}\right]_0^{1-x} dx = \int_0^1 \left[x^2(1-x) + \frac{(1-x)^3}{3}\right] dx$$

$$= \left[\frac{x^3}{3} - \frac{x^4}{4} - \frac{(1-x)^4}{3\times 4}\right]_0^1 = \left[\frac{1}{3} - \frac{1}{4} + \frac{1}{12}\right] = \frac{1}{6}.$$

Example 33:

Evaluate $\iint_A (x^2 + y^2)\,dx\,dy$, *where A is the region bound by* $x = 0$, $y = 0$, $x + y = 1$.

Solution:

Do your self.

The region A is bounded by y = 0, y = 1 – x and x = 0, x = 1.

Example 34:

Evaluate $\iint xy(x + y)\,dx\,dy$ *over the area between* $y = x^2$ *and* $y = x$.

Solution:

Draw the given curves $y = x^2$ and y = x in the same figuare. The two curves intersect at the points whose abscissae are given by $x^2 = x$ or x (x – 1) = 0 *i.e.*, x = 0 or 1. When 0 < x < 1, we have $x > x^2$. So the area of integration can be considered as lying between the curves $y = x^2$, y = x, x = 0 and x = 0 and x = 1.

Therefore the required integral

$$= \int_{x=0}^{1}\int_{y=x^2}^{x} xy(x+y)\,dx\,dy = \int_0^1\left[\int_{x^2}^{x}(x^2y+xy^2)xy\right]dx$$

$$= \int_0^1\left[\frac{x^2y^2}{2}+\frac{xy^3}{3}\right]_{x^2}^{x} dx = \int_0^1\left[\left(\frac{x^4}{2}+\frac{x^4}{3}\right)-\left(\frac{x^6}{2}+\frac{x^7}{3}\right)\right]dx$$

$$= \int_0^1\left[\frac{5x^4}{6}-\frac{x^6}{2}-\frac{x^7}{3}\right]dx = \left[\frac{x^5}{6}-\frac{x^7}{14}-\frac{x^8}{24}\right]_0^1$$

$$= \frac{1}{6}-\frac{1}{14}-\frac{1}{24} = \frac{28-12-7}{168} = \frac{9}{168} = \frac{3}{56}.$$

Example 35:

Find by double integration the area lying between the parabola $y = 4x - x^2$ and the line $y = x$.

Solution:

Solving $y = 4x - x^2$ and $y = x$ for x, we have $4x - x^2$ or $x^2 - 3x = 0$ or $x(x - 3) = 0$ *i.e.*, $x = 0$ or 3.

This the curves $y = 4x - x^2$ and $y = x$ intersect at the points where $x = 0$ and $x = 3$. When $0 < x < 3$, we have $4x - x^2 > x$.

So the required area can be considered as lying between the curves $y = x$, $y = 4x - x^2$, $x = 0$ and $x = 3$.

Therefore the required area $= \int_{x=0}^{3}\int_{y=x}^{4x-x^2} dx\,dy$

$$= \int_0^3 [y]_x^{4x-x^2} dx = \int_0^3 (4x-x^2-x)dx = \int_0^3 (3x-x^2)dx$$

$$= \left[\frac{3x^2}{2}-\frac{x^3}{3}\right]_0^3 = \frac{27}{2}-\frac{27}{3} = 27\cdot\frac{1}{6} = \frac{9}{2}.$$

Example 36:

Prove by the method of double integration that the area lying between the parabolas $y^2 = 4ax$ *and* $x^2 = 4ay$ *is* $\frac{16}{3}a^2$

Solution:

Draw the two parabolas in the same figure. The two parabolas intersect at the points whose abscissase are given by $(x^2/4a)^2 = 4ax$ *i.e.*, $x(x^3 - 64a^3) = 0$ *i.e.*, $x = 0$ and $x^3 = 64a^3$. Thus the two parabolas intersect at the points where $x = 0$ and $x = 4a$.

Now the area of a small element situated at any point (x,y) = dx dy.

$\therefore$ the required area $= \int_{x=0}^{4a}\int_{y=x^2/4a}^{\sqrt{(4ax)}} dx\,dy = \int_0^{4a}[y]_{x^2/4a}^{\sqrt{(4ax)}} dx$

$$= \int_0^{4a}\left[2\sqrt{a}.x^{1/2} - \frac{1}{4a}.x^2\right]dx = \left[2\sqrt{a}.x^{3/2}.\frac{2}{3} - \frac{1}{4a}\cdot\frac{x^3}{3}\right]_0^{4a}$$

$$= \frac{4}{3}\sqrt{a}.(4a)^{3/2} - \frac{1}{12a}.64a^3 = \frac{32}{3}a^2 - \frac{16}{3}a^2 = \frac{16}{3}a^2.$$

Example 37:

Evaluate $\iint y\,dx\,dy$ *over the area between the parabolas* $y^2 = 4x$ *and* $x^2 = 4y$.

Solution:

The two parabolas intersect at the points whose abscissae area given by $\left(\frac{1}{4}x^2\right)^2 = 4x$ or $x(x^3 - 64) = 0$ *i.e.*, x = 0 or 4. When 0 < x < 4, we have $2\sqrt{x} > \frac{1}{4}x^2$. Therefore the given region of integration can be expressed as $0 \le x \le 4$, $\frac{1}{4}x^2 \le y \le 2\sqrt{x}$.

$\therefore$ the required integral $= \int_{x=0}^{4}\int_{y=x^2/4}^{2\sqrt{x}} y\,dx\,dy$

$$= \int_0^4\left[\frac{y^2}{2}\right]_{x^2/4}^{2\sqrt{x}} dx = \int_0^4\left[2x - \frac{x^4}{32}\right]dx = \left[\frac{2x^2}{2} - \frac{x^5}{32\times 2}\right]_0^4$$

$$= 16 - \frac{32}{5} = \frac{48}{5}.$$

Example 38:

When the region of integration A is the triangle given by y = 0, y = x and x = 1, show that

$$\iint_A \sqrt{(4x^2 - y^2}(dx\,dy) = \frac{1}{3}\left(\frac{\pi}{3} + \frac{\sqrt{3}}{2}\right)$$

Solution:

In the diagram draw the straight lines y = 0, y = x and x = 1. Then we odserve that the region of integration A can be expressed as $0 \le y \le x$, $0 \le x \le 1$.

$$\therefore \iint_A \sqrt{(4x^2 - y^2)}\,dx\,dy = \int_{x=0}^{1}\int_{y=0}^{x} \sqrt{(4x^2 - y^2)}\,dx\,dy$$

$$= \int_0^1 \left[\frac{y}{2}\sqrt{(4x^2 - y^2)} + 2x^2 \sin^{-1}\frac{y}{2x}\right]_{y=0}^{x} dx,$$

integrating w.r.t. y treating x as constant

$$= \int_0^1 \left[\frac{x}{2}\sqrt{(4x^2 - y^2)} + 2x^2 \sin^{-1}\frac{1}{2} - 0\right] dx$$

$$= \int_0^1 \left[\frac{\sqrt{3}}{2}x^2 + \frac{\pi}{3}x^2\right] dx = \left[\frac{\sqrt{3}}{2}\cdot\frac{x^3}{3} + \frac{\pi}{3}\cdot\frac{x^3}{3}\right]_0^1$$

$$= \frac{1}{3}\left(\frac{\sqrt{3}}{2} + \frac{\pi}{3}\right).$$

Example 39:

Evaluate $\iint \frac{xy}{\sqrt{(1-y^2)}}\,dx\,dy$ *over the positive quadrant of the circle* $x^2 + y^2 = 1$.

Solution:

Here the region of integration R is she area of the circle $x^2 + y^2 =$ lying in the positive quadrant. This region of integration R can be expressed as

$0 \le x \le \sqrt{(1 - y^2)},\ 0 \le y \le 1.$

$$\therefore \iint_R \frac{dy}{\sqrt{(1-y^2)}}\,dx\,dy$$

$$= \int_{y=0}^{1}\int_{x=0}^{\sqrt{(1-y^2)}} \frac{xy}{\sqrt{(1-y^2)}}\,dx\,dy \quad \int_0^1 \frac{y}{\sqrt{(1-y^2)}}\left[\frac{x^2}{2}\right]_{x=0}^{\sqrt{(1-y^2)}} dy,$$

integrating w.r.t. x treating y as constant

$$= \frac{1}{2}\int_0^1 y\sqrt{(1-y^2)}\,dy = \frac{1}{2}\int_0^1 -\frac{1}{2}.(1-y^2)^{1/2}(-2y)\,dy$$

$$= -\frac{1}{4}\cdot\frac{2}{3}\left[(1-y^2)^{3/2}\right]_0^1, \text{ by power formula}$$

$$= \frac{1}{6}.$$

Example 40:

Evaluate the double integral $\int_0^a \int_0^{\sqrt{(a^2-x^2)}} x^2 y\,dx\,dy.$

Mention the region of integration involved in this double integral.

Solution:

The given integral

$$I = \int_{x=0}^{a}\int_{y=0}^{\sqrt{(a^2-x^2)}} x^2 y \, dx\, dy$$

$$= \int_0^a x^2 \left[\frac{y^2}{2}\right]_{y=0}^{\sqrt{(a^2-x^2)}} dx, \text{integrating w.r.t. y treating x as constant}$$

$$= \frac{1}{2}\int_0^a x^2(a^2-x^2)dx = \frac{1}{2}\int_0^a (a^2x^2 - x^4)dx$$

$$= \frac{1}{2}\left[a^2\frac{x^3}{3} - \frac{x^5}{5}\right]_0^a = \frac{1}{2}\left[\frac{a^5}{3} - \frac{a^5}{5}\right] = \frac{1}{15}a^5.$$

From the limits of integration it is obvious that the region of integration R is bounded by $y = 0$, $y = \sqrt{(a^2 - x^2)}$ and $x = 0$, $x = a$ *i.e.*, the region of integration is the area of the circle $x^2 + y^2 = a^2$ between the lines $x = 0$, $x = a$ and lying above the line $y = 0$ *i.e.*, the axis of x. Thus the region of integration is the area OAB of the circle $x^2 + y^2 = a^2$ lying in the positive quadrant.

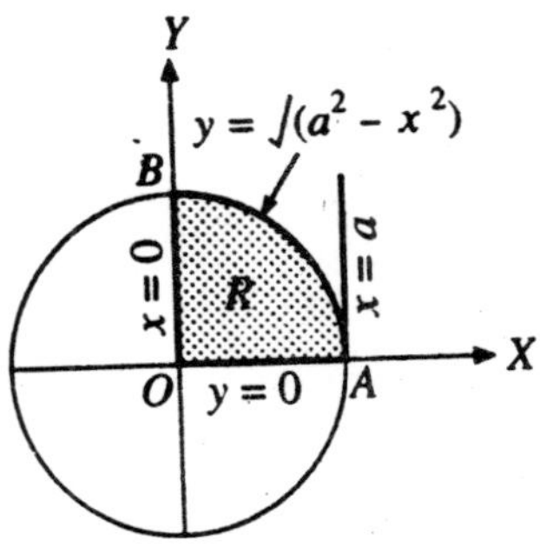

Fig. 1.5

Example 41:

Evaluate $\int_0^{\pi}\int_0^{a\sin\theta} r\, d\theta\, dr$.

Solution:

Here the limits of r are variable and those of θ are constant. Therefore first integration shall be performed w.r.t. r tegarding θ as a constant. We have

$$\int_0^{\pi}\int_0^{a\sin\theta} r\, d\theta\, dr = \int_0^{\pi}\left[\frac{r^2}{2}\right]_0^{a\sin\theta} d\theta = \frac{1}{2}\int_0^{\pi} a^2\sin^2\theta\, d\theta$$

$$= \frac{a^2}{2}.2\int_0^{\pi/2}\sin^2\theta\, d\theta = \frac{a^2}{2}\cdot 2\cdot\frac{1}{2}\cdot\frac{\pi}{2} = \frac{\pi a^2}{4}.$$

Example 42:

Evaluate $\int_0^{\pi/2}\int_0^{a\cos\theta} r\sin\theta\, d\theta\, dr$.

Solution:

We have $\int_0^{\pi/2}\int_0^{a\cos\theta} r\sin\theta\, d\theta\, dr = \int_0^{\pi/2}\sin\theta\left[\frac{r^2}{2}\right]_0^{a\cos\theta} d\theta$'

integrating first w.r.t. r regarding θ as a constant

$$= \frac{1}{2}\int_0^{\pi/2}\sin\theta . a^2\cos^2\theta\, d\theta = \frac{a^2}{2}\int_0^{\pi}\sin\theta\cos^2\theta\, d\theta$$

$$= \frac{1}{2}a^2 . \frac{1.1}{3.1} = \frac{1}{6}a^2.$$

Example 43:

Evaluate $\int_0^{\pi}\int_0^{a(1+\cos\theta)} r^2\cos\theta\, d\theta\, dr$.

Solution:

We have $\int_0^{\pi}\int_0^{a(1+\cos\theta)} r^2\cos\theta\, d\theta\, dr = \int_0^{\pi}\cos\theta\left[\frac{r^3}{3}\right]_0^{a(1+\cos\theta)} d\theta$

$$= \frac{1}{3}\int_0^{a}\cos\theta . a^2(1+\cos\theta)^3 d\theta$$

$$= \frac{a^3}{3}\int_0^{\pi}\cos\theta(1+3\cos\theta+3\cos^2\theta+\cos^3\theta)d\theta$$

$$= \frac{a^3}{3}\int_0^{\pi}[\cos\theta+3\cos^2\theta+\cos^4\theta]d\theta$$

$$= 2.\frac{a^2}{3}\int_0^{\pi/2}[3\cos^2\theta+\cos^4\theta]d\theta$$

$$\left[\because \int_0^{\pi}\cos^n\theta\, d\theta = 0 \text{ or } 2\int_0^{\pi/2}\cos^n\theta\, d\theta \text{ according as n is odd or even}\right]$$

$$= \frac{2a^3}{3}\left[3\cdot\frac{1}{2}\cdot\frac{\pi}{2}+\frac{3.1}{4.2}\cdot\frac{\pi}{2}\right] = \frac{2a^2}{3}\cdot\frac{3\pi}{4}\left[1+\frac{1}{4}\right]$$

$$= \frac{2a^3}{3}\cdot\frac{3\pi}{4}\cdot\frac{5}{4} = \frac{5\pi a^3}{8}.$$

Example 44:

Evaluate the triple integral of the function $f(x, y, z) = x^2$ *over the region* V *enclosed by the planes* $x = 0$, $y = 0$, $z = 0$ *and* $x + y + z = a$.

Solution:

The given region V is bounded by the co-ordinate planes x = 0, y = 0, z = 0 and the plane x + y + z = a. To cover the region V, let the values of x, y lie within the triangle bounded by x-axis, the y-axis and the line (x + y = a, z = 0). Then for any point (x, y, 0) within this triangle, z varies from z = a − x − y in the region V.

But the values of x and y vary within the triangle formed in the xy-plane. Therefore x varies from 0 to a and for any intermediary value of x, y varies from 0 to a − x.

Therefore, the region of integration V can be expressed as $0 \le x \le a$, $0 \le y \le a - x$, $0 \le z \le a - x - y$.

Hence the required triple integral

$$= \int_0^a \int_0^{a-x} \int_0^{a-x-y} x^2 dx\, dy\, dz$$

Example 45:

Find the volume of the tetrahedron bounded by the coordinate planes and the plane x + y + z = 1.

Solution:

Here the region of integration V to cover the volume of the tetrahedron can be expressed as $0 \le x \le 1$, $0 \le y \le 1 - x$, $0 \le z \le 1 - x - y$.

Therefore the required volume of the tetrahedron

$$= \iiint_V dx\, dy\, dz = \int_0^1 \int_0^{1-x} \int_0^{1-x-y} dx\, dy\, dz$$

$$= \int_0^1 \int_0^{1-x} [z]_0^{1-x-y} dx\, dy = \int_0^1 \int_0^{1-x} (1 - x - y) dx\, dy$$

$$= \int_0^1 \left[(1-x)y - \frac{y^2}{2}\right]_0^{1-x} dx = \int_0^1 \left[(1-x)^2 - \frac{(1-x)^2}{2}\right] dx$$

$$= \int_0^1 \frac{1}{2}(1-x)^2 dx = \frac{1}{2}\left[\frac{(1-x)^3}{3.(-1)}\right]_0^1 = -\frac{1}{6}[0-1] = \frac{1}{6}.$$

Example 46:

Find the volume of the tetrahedron bounded by the plane x/a + y/b + z/c = 1 *and the coordinate planes.*

Solution:

Here the region of integration V to cover the volume of the given tetrahedron can be depressed as $0 \le x \le a$, $0 \le y \le b(1 - x/a)$, $0 \le z \le c(1 - x/a - y/b)$.

Therefore the required volume of the tetrahedron

$$= \iiint_V dx\,dy\,dz = \int_0^a \int_0^{b(1-x/a)} \int_0^{c(1-x/a-y/b)} dx\,dy\,dz.$$

The required volume $= \dfrac{abc}{6}$.

Example 47:

Find the volume of a sphere of radius a by triple integral.

Solution:

Referred to centre as origin the equation of a sphere of radius a is $x^2 + y^2 + z^2 = a^2$. ...(1)

The sphere (1) symmetrical in all the eight octants.

$\therefore$ volume of the sphere (1) = 8. (the volume of the part of the sphere lying in the positive octant).

Now for the region consisting of the volume of the sphere (1) lying in the positive octant, we have $0 \le x \le a$, $0 \le y \le \sqrt{(a^2 - x^2)}$, $0 \le z \le \sqrt{(a^2 - x^2 - y^2)}$.

$\therefore$ the required volume of a sphere of radius a

$$= 8\int_{x=0}^{a} \int_{y=0}^{\sqrt{(a^2-x^2)}} \int_{z=0}^{\sqrt{(a^2-x^2-y^2)}} dx\,dy\,dz$$

$$= 8\int_{x=0}^{a} \int_{y=0}^{\sqrt{(a^2-x^2)}} \left[z\right]_{z=0}^{\sqrt{(a^2-x^2-y^2)}} dx\,dy$$

$$= 8\int_{x=0}^{a} \int_{y=0}^{\sqrt{(a^2-x^2)}} \sqrt{(a^2 - x^2 - y^2)}\,dx\,dy$$

$$= 8\int_0^a \left[\frac{y}{2}\sqrt{\{(a^2 - x^2) - y^2\}} + \frac{a^2 - x^2}{2}\sin^{-1}\frac{y}{\sqrt{(a^2 - x^2)}}\right]_{y=0}^{\sqrt{(a^2-x^2)}} dx$$

$$= 8\int_0^a \left[0 + \frac{a^2 - x^2}{2}\cdot\frac{\pi}{2} - 0 - 0\right] dx$$

$$= 8\cdot\frac{\pi}{4}\int_0^a (a^2 - x^2)\,dx = 2\pi\left[a^2 x - \frac{x^3}{3}\right]_0^a = 2\pi\left[a^3 - \frac{a^3}{3}\right] = \frac{4}{3}\pi a^3.$$

Example 48:

Evaluate $\iiint (x + y + z)\,dx\,dy\,dz$ *over the tetrahedron* $x = 0$, $y = 0$, $z = 0$ *and* $x + y + z = 1$.

Solution:

The region of integrating V for the given tetrahedron can be expressed as $0 \leq x \leq 1$, $0 \leq y \leq 1-x$, $0 \leq z \leq 1-x-y$.

Hence the required triple integral $= \iiint_V (x+y+z)\,dx\,dy\,dz$

$$= \int_0^1 \int_0^{1-x} \int_0^{1-x-y} (x+y+z)\,dx\,dy\,dz$$

$$= \int_0^1 \int_0^{1-x} \left[(x+y)z + \frac{z^2}{2}\right]_0^{1-x-y} dx\,dy$$

$$= \int_0^1 \int_0^{1-x} \left[(x+y)(1-x-y) + \frac{(1-x-y)^2}{2}\right] dx\,dy$$

$$= \int_0^1 \int_0^{1-x} (1-x-y)\left(x+y+\frac{1-x-y}{2}\right) dx\,dy$$

$$= \int_0^1 \int_0^{1-x} \frac{1}{2}(1-x-y)(1+x+y)\,dx\,dy$$

$$= \frac{1}{2}\int_0^1 \int_0^{1-x} [1-(x+y)^2]\,dx\,dy = \frac{1}{2}\int_0^1 \left[y - \frac{(x+y)^3}{3}\right]_0^{1-x} dx$$

$$= \frac{1}{2}\int_0^1 \left(1 - x - \frac{1}{3} + \frac{x^3}{3}\right) dx = \frac{1}{2}\int_0^1 \left(\frac{2}{3} - x + \frac{x^3}{3}\right) dx$$

$$= \frac{1}{2}\left[\frac{2}{3}x - \frac{x^2}{2} + \frac{x^4}{3\times 4}\right]_0^1 = \frac{1}{2}\left[\frac{2}{3} - \frac{1}{2} + \frac{1}{12}\right] = \frac{1}{2}\cdot\frac{1}{4}\cdot = \frac{1}{8}.$$

Example 49:

Evaluate the integral $\iiint xyz\,dx\,dy\,dz$ *over the volume enclosed by three coordinate planes and the plane* $x + y + z = 1$.

Solution:

The region of integration V enclosed by the three coordinate planes and the plane $x + y + z = 1$ can be expressed as $0 \leq x \leq 1$, $0 \leq y \leq 1-x$, $0 \leq z \leq 1-x-y$.

$\therefore$ the required triple integral $\iiint_V xyz\,dx\,dy\,dz$

$$= \int_{x=0}^{1} \int_{y=0}^{1-x} \int_{z=0}^{1-x-y} xyz\,dx\,dy\,dz.$$

Example 50:

Evaluate $\iiint \frac{dx\,dy\,dz}{(x+y+z+1)^3}$ *over the reagin* $x \geq 0,\ y \geq 0,\ z \geq 0,\ x+y+z \leq 1$.

Solution:

The given region of integration R can be expressed as $0 \leq x \leq 1,\ 0 \leq y \leq 1-x,\ 0 \leq z \leq 1-x-y$.

Hence the required triple integral

$$= \iiint_R \frac{dx\,dy\,dz}{(x+y+z+1)^3}$$

$$= \int_0^1 \int_0^{1-x} \int_0^{1-x-y} \frac{1}{(x+y+z+1)^3}\,dx\,dy\,dz$$

$$= \int_0^1 \int_0^{1-x} \left[\int_0^{1-x-y} (x+y+z+1)^{-3} dz \right] dx\,dy$$

$$= \int_0^1 \int_0^{1-x} \left[\frac{(x+y+z+1)^{-2}}{-2} \right]_0^{1-x-y} dx\,dy$$

$$= -\frac{1}{2} \int_0^1 \int_0^{1-x} \left[\frac{1}{4} - \frac{1}{(x+y+1)^2} \right] dx\,dy$$

$$= -\frac{1}{2} \int_0^1 \left[\frac{y}{4} + \frac{1}{(x+y+1)} \right]_0^{1-x} dx$$

$$= -\frac{1}{2} \int_0^1 \left[\frac{1-x}{4} + \frac{1}{2} - \frac{1}{(x+1)} \right] dx$$

$$= -\frac{1}{2} \left[\frac{(1-x)^2}{2 \times 4 \times (-1)} + \frac{1}{2}x - \log(x+1) \right]_0^1$$

$$= -\frac{1}{2} \left[\left\{ 0 + \frac{1}{2} - \log 2 \right\} - \left\{ -\frac{1}{8} + 0 - 0 \right\} \right] = \frac{1}{2} \left[\frac{1}{2} - \log 2 + \frac{1}{8} \right]$$

$$= -\frac{1}{2} \left[\frac{5}{8} - \log 2 \right] = \frac{1}{2} \left[\log 2 - \frac{5}{8} \right].$$

Example 51:

Evaluate $\iiint xyz\,dx\,dy\,dz$ *over the ellipsoid* $x^2/a^2 + y^2/b^2 + z^2/c^2 = 1$.

Solution:

Hence the region of integration can be expressed as $-a \le x \le a$, $-b\sqrt{\{1-(x^2/a^2)\}} \le y \le b\sqrt{\{1-(x^2/a^2)\}}$ and $-c\sqrt{\{1-(x^2/a^2)-(y^2/b^2)\}} \le z \le c\sqrt{\{1-(x^2/a^2)-(y^2/b^2)\}}$.

$\therefore$ the required triple integral

$$= \int_{-a}^{a}\int_{-b\sqrt{\{1-(x^2/a^2)\}}}^{b\sqrt{\{1-(x^2/a^2)\}}}\left[\int_{-c\sqrt{\{1-(x^2/a^2)-(y^2/b^2)\}}}^{c\sqrt{\{1-(x^2/a^2)-(y^2/b^2)\}}}(xy).z\,dz\right]dx\,dy$$

$= 0$, [$\because$ z is an odd function of z and xy is treated as constant while integrating w.r.t. z]

Example 51:

Evaluate $\iiint z^2\,dx\,dy\,dz$ *over the sphere* $x^2 + y^2 + z^2 = 1$.

Solution:

Here the region of integration can be expressed as

$-1 \le x \le 1$, $-\sqrt{(1-x^2)} \le y \le \sqrt{(1-x^2)}$,

$-\sqrt{(1-x^2-y^2)} \le z \le \sqrt{(1-x^2-y^2)}$.

$\therefore$ the required triple integral

$$= \int_{-1}^{1}\int_{-\sqrt{(1-x^2)}}^{\sqrt{(1-x^2)}}\int_{-\sqrt{(1-x^2-y^2)}}^{\sqrt{(1-x^2-y^2)}} z^2\,dx\,dy\,dz$$

$$= \int_{-1}^{1}\int_{-\sqrt{(1-x^2)}}^{\sqrt{(1-x^2)}}\left[\frac{z^3}{3}\right]_{-\sqrt{(1-x^2-y^2)}}^{\sqrt{(1-x^2-y^2)}} dx\,dy$$

$$= \frac{1}{3}\int_{-1}^{1}\left[\int_{-\sqrt{(1-x^2)}}^{\sqrt{(1-x^2)}} 2(1-x^2-y^2)^{3/2}\,dy\right]dx$$

$$= \frac{2}{3}\int_{-1}^{1}\left[\int_{-\pi/2}^{\pi/2}[(1-x^2)\cos^2\theta]^{3/2}.\sqrt{(1-x^2)}.\cos\theta\,d\theta\right]dx$$

[putting $y = \sqrt{(1-x^2)}\sin\theta$ so that $dy = \sqrt{(1-x^2)}\cos\theta\,d\theta$; also when $y = 0$, $\theta = 0$ and when $y = \sqrt{(1-x^2)}$, $\theta = \pi/2$]

$$= \frac{2}{3}\int_{-1}^{1}\left[2.\int_{0}^{\pi/2}(1-x^2)^2\cos^4\theta\,d\theta\right]dx$$

$$= \frac{4}{3}\int_{-1}^{1}(1-x^2)^2\cdot\frac{3.1}{4.2}\cdot\frac{\pi}{2}dx = \frac{\pi}{2}\int_{-1}^{1}(1-x^2)^2\,dx$$

$$= \frac{\pi}{4}\cdot 2\int_{0}^{1}(1-2x^2+x^4)dx = \frac{\pi}{2}\left[x-\frac{2}{3}x^3+\frac{1}{5}x^5\right]_{0}^{1}$$

$$= \frac{\pi}{2}\left[1 - \frac{2}{3} + \frac{1}{5}\right] = \frac{\pi}{2}\cdot\frac{8}{15} = \frac{4\pi}{15}.$$

Example 52:

Evaluate $\iiint (z^5 + z)\,dx\,dy\,dz$ *over the sphere* $x^2 + y^2 + z^2 = 1$.

Solution:

The given region of integration can be expressed as

$-1 \le x \le 1,\ -\sqrt{(1 - x^2)} \le y \le \sqrt{(1 - x^2)}$,

$-\sqrt{(1 - x^2 - y^2)} \le z \le \sqrt{(1 - x^2 - y^2)}$.

Hence the required triple integral

$$= \int_{-1}^{1}\int_{-\sqrt{(1-x^2)}}^{\sqrt{(1-x^2)}}\left[\int_{-\sqrt{(1-x^2-y^2)}}^{\sqrt{(1-x^2-y^2)}} (z^5 + z)\,dz\right]dx\,dy$$

$= 0,$ $\qquad$ [$\because$ $(z^5 + z)$ is an odd function of z].

Example 53:

Evaluate $\iiint_R u^2 v^2 w\,du\,dv\,dw$, *where R is the region* $u^2 + v^2 \le 1,\ 0 \le w \le 1$.

Solution:

Here the limits of integration to cover the region R can be taken as $-1 \le u \le 1,\ -\sqrt{(1 - u^2)} \le v \le \sqrt{(1 - u^2)},\ 0 \le w \le 1$, where the first integration is to be performed with respect to v.

$$\therefore \iiint_R u^2 v^2 w\,du\,dv\,dw = \int_0^1\int_{-1}^{1}\int_{-\sqrt{(1-u^2)}}^{\sqrt{(1-u^2)}} u^2 v^2 w\,dw\,du\,dv$$

$$= \int_0^1\int_{-1}^{1} u^2 w\left[\int_{-\sqrt{(1-u^2)}}^{\sqrt{(1-u^2)}} v^2\,dv\right]dw\,du,$$

because the first integration is to be performed w.r.t. v regarding u and w as constants

$$= \int_0^1\int_{-1}^{1}\left[2u^2 w\int_0^{\sqrt{(1-u^2)}} v^2\,dv\right]dw\,du$$

because v^2 is an even function of v

$$= \int_0^1\int_{-1}^{1} 2u^2 w\left[\frac{v^3}{3}\right]_0^{\sqrt{(1-u^2)}} dw\,du$$

$$= \frac{2}{3}\int_0^1\int_{-1}^{1} wu^2 (1 - u^2)^{3/2}\,dw\,du$$

$$= \frac{2}{3}\int_0^1\left[w.2\int_0^1 u^2(1-u^2)^{3/2}\,du\right]dw$$

$$= \frac{4}{3}\int_0^1 w\left[\int_0^{\pi/2}\sin^2\theta.\cos^3.\cos\theta\,d\theta\right]dw, \text{ putting } u = \sin\theta$$

$$= \frac{4}{3}\int_0^1 w\left[\int_0^{\pi/2}\sin^2\theta\cos^4\theta\,d\theta\right]dw = \frac{4}{3}\int_0^1 w\cdot\frac{1.3.1}{6.4.2}\cdot\frac{\pi}{2}dw$$

$$= \frac{\pi}{24}\int_0^1 w\,dw = \frac{\pi}{24}\left[\frac{w^2}{2}\right]_0^1 = \frac{\pi}{48}[1-0] = \frac{\pi}{48}.$$

Example 54:

Change the order of integration in the double integral $\int_0^a\int_0^x f(x,y)dx\,dy$.

Solution:

In the given integral the limits if integration are given by the straight lines y = 0, y = x = 0 and x = a. Draw these lines bounding the region of integration in the same figure. We observe the the region of integration is the area ONM.

In the given integral, the limits of integration of y being variable, we are required to integrate first w.r.t. y regarding x as constant and then w.r.t. x.

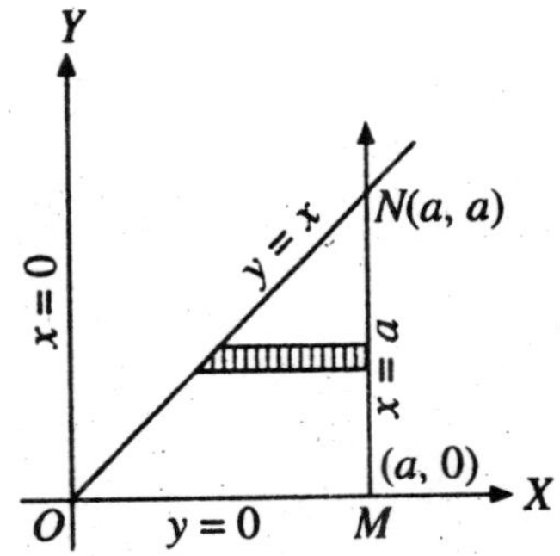

Fig. 1.6

To reverse the order of integration, we have to integrate first w.r.t. x regarding y as constant and then w.r.t. y. This is done by dividing the area ONM into strips parallel to the x-axis. Let us take strips parallel to the x-axis starting from the line ON (*i.e.*, y = x) and terminating on the line MN (*i.e.*, x = a). Thus for this region ONM, x varies from y to a and y varies from 0 to a.

Hence the changing the order of integration, we have

$$\int_0^a\int_0^x f(x,y)dx\,dy = \int_0^a\int_y^a f(x,y)dx\,dy.$$

Example 55:

Prove that $\int_a^b dx\int_a^x f(x,y)dy = \int_a^b dy\int_y^b f(x,y)dx$.

Solution:

Let $I = \int_a^b dx \int_a^x f(x,y)dy$.

We are required to change the order of integration in the integral I. In the integral I the limits of integration of y are given by the straight lines y = a and y = x. Also the limits of integration of x are given by the straight lines x = a and x = b. Draw the straight lines y = a, y = x, x = a and x = b, bounding the region of integration, in the same figure. We observe that the region of integration is the area of the triangle ABC.

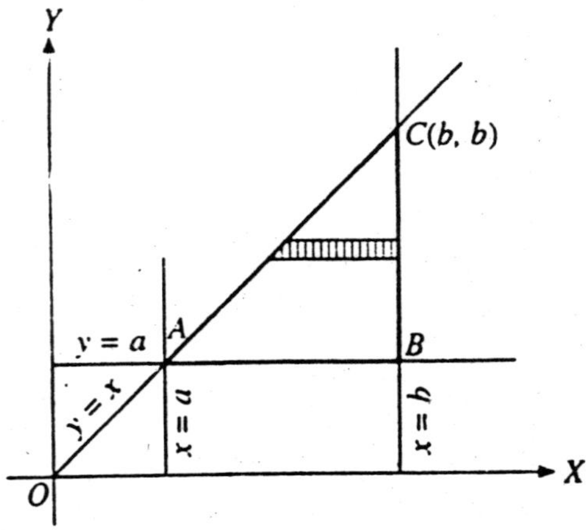

Fig. 1.7

In the integral I we are required to integrate first w.r.t. y and then w.r.t. x. To reverse the order of integration we have to integrate first w.r.t. x and then w.r.t. y. This is done by dividing the area ABC into strips parallel to the x-axis. Let us take strips parallel to the x-axis starting from the line AC (*i.e.*, y = x) and terminating on the line BC (*i.e.*, x = b). Thus for the region ABC, x-varies from y to b and y varies from a to b. Hence by changing the order of integration, we have

$$\int_a^b dx \int_a^x f(x,y)dy = \int_a^b dy \int_y^b f(x,y)dx.$$

Example 56:

Change the order of integration in $\int_0^1 \int_x^{x(2-x)} f(x,y)dx\,dy$.

Solution:

In the given integral the limits of integration of y are given by y = x, which is a straight line passing through the origin, and y = x (2 – x) or $y = 2x - x^2$ or $(x - 1)^2 = -(y - 1)$ which is a parabola with vertex (1, 1) a and passing through the origin.

Again the limits of integration of x are given by x = 0 *i.e.*, the y-axis and x = 1 which is a straight line parallels to the y-axis at a distance 1 from the origin.

We draw the curves y = x, $(x - 1)^2 = -(y - 1)$, x = 0 and x = 1, giving the limits of integration, in the same figure. We observe that the region of integration is the area OLBMO.

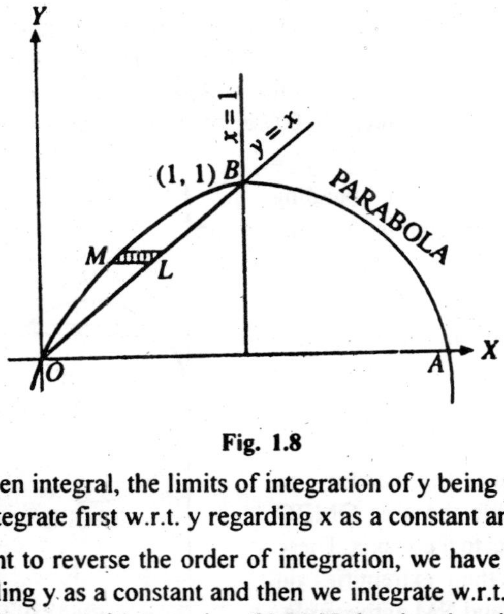

Fig. 1.8

In the given integral, the limits of integration of y being variable, we are required to integrate first w.r.t. y regarding x as a constant and then w.r.t. x.

If we want to reverse the order of integration, we have to fist integrate w.r.t. x regarding y as a constant and then we integrate w.r.t. y. This is done by covering the area of integration OLBMO by drawing the straight lines y = constant *i.e.*, by dividing this area into strips parallel to the x-axis.

So divide the region OLBMO into strips parallel to the x-axis starting from the arc OMB of the parabola and terminating on the line OLB.

For the point B, x = 1. Putting x = 1 in the equation of the line y = x, we get y = 1. So the y-coordinate of the point B is also 1.

For the region OMBLO, the lower limit of x is the value of x found interns of y from the equation $(x - 1)^2 = 1 - y$ and the upper limit of x is the value of x found in terms of y from the equation $y = x$. From the equation $(x-1)^2 = 1 - y$, we get $x - 1 = \pm \sqrt{(1-y)}$ or $x = 1 \pm \sqrt{(1-y)}$. Since in the region OMBLO, x takes values less than 1, therefore we take $x = 1 - \sqrt{(1-y)}$.

Thus in the region OMBLO, x varies form $1 - \sqrt{(1-y)}$ to y and y varies from 0 to a.

Hence by changing the order of integration, we have the given integral

$$= \int_0^1 \int_{1-\sqrt{(1-y)}}^{y} f(x,y)\,dy\,dx.$$

Example 57:

Change the order of integration in $\int_0^{2a} \int_0^{\sqrt{(2ax-x^2)}} f(x,y)\,dx\,dy.$

Solution:

In the given integral the limits of integration of are given by $y = 0$ (*i.e.*, the x-axis) and $y = \sqrt{(2ax - x^2)}$ *i.e.*, $y^2 = 2ax - x^2$ *i.e.*, $(x - a)^2 + y^2 = a^2$ which is a circle with centre (a, 0) and raidius a. Again the limits of integration of x are given by the straight lines $x = 0$ (*i.e.*, the y-axis) and $x = 2a$.

Draw the curves $(x + a)^2 + y^2 = a^2$, $y = 0$, $x = 0$ and $x = 2a$, bounding the region of integration, in the same figure. From figure we observe that the area of integration is OMNO.

In the given integral we are required to integrate first w.r.t y regarding x as a constant ans then w.r.t. x.

To reverse the order of integration, divide the area OMBO into strips parallel to the x-axis. These strips will have their extremities on the portions ON and NM of the circle.

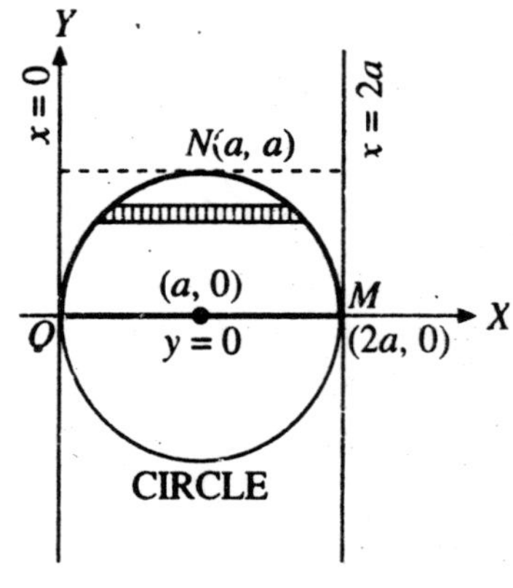

Fig. 1.9

Solving the equation of circle $(x - a)^2 + y^2 = a^2$ for x, we get $(x - a)^2 = a^2 - y^2$ *i.e.*, $x - a = \pm \sqrt{(a^2 - y^2)}$ *i.e.*, $x = a \pm \sqrt{(a^2 - y^2)}$.

So for the region OMNO, x varies from $a - \sqrt{(a^2 - y^2)}$ to $a + \sqrt{(a^2 - y^2)}$ and y varies from 0 to a.

Therefore, changing the order of integration, the given double integral transforms to $\int_0^a \int_{a-\sqrt{(a^2-y^2)}}^{a+\sqrt{(a^2-y^2)}} f(x,y)\,dy\,dx.$

Example 58:

Change the order of integration in the integral

$$\int_0^3 \int_0^{\sqrt{(4-y)}} (x+y)\,dy\,dx.$$

Solution:

In the given integral the limits of integration of x are given by the straight line x= 1 and the curve $x = \sqrt{(4-y)}$ *i.e.*, $x^2 = 4 - y$ *i.e.*, $x^2 = -(y-4)$ which is a parabola, symmetrical about the y-axis, with vertex at the point (0, 4) and existing in the region $y \leq 4$. Again the limits of integration of y are given by the straight lines y = 0 (*i.e.*, the x-axis) and y = 3.

We draw the curves x = 1, $x^2 = -(y-4)$, y = 0 and y = 3, giving the limits of integration in the same figure. Putting x = 1 in the equation $x^2 = -(y-4)$, we get y = 3. Thus the straight line y = 3 passes through the point of intersection C of x = 1 and $x^2 = -(y-4)$. Also at the point of intersection B of the parabola $x^2 = -(y-4)$ and the x-axis (*i.e.*, the line y = 0), we have x = 2. We observe that the region of integration is the area ABCA.

In the given integral the limits of integration of x are variable while those of y are constant. Thus we have to first integrate w.r.t. x regarding y as a constant and then we integrate w.r.t. y.

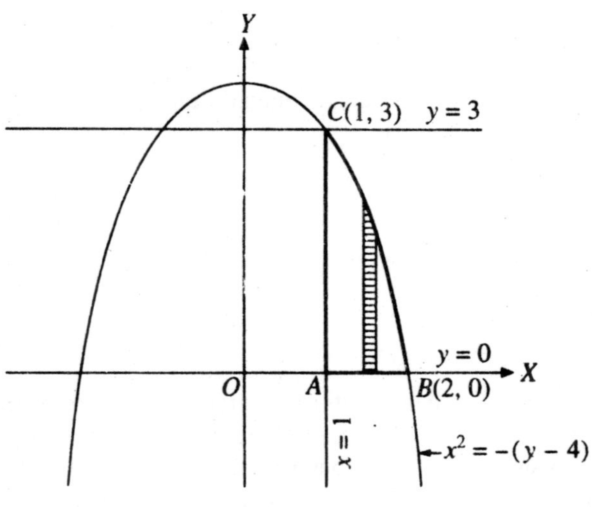

Fig. 1.10

If we want to change the order of integration, we have to first integrate w.r.t. y regarding x as a constant and then we integrate w.r.t. x. This is done by covering the area ABCA by strips drawn parallel to the y-axis. These strips start from the line AB (*i.e.*, $y = 0$) and terminate on the arc BC of the parabola $x^2 = 4 - y$. Therefore, for the region ABCA, y varies from 0 to $4 - x^2$ and x varies from 1 to 2. Hence by changing the order of integration, we have the given integral

$$= \int_1^2 \int_0^{4-x^2} (x+y)\, dx\, dy$$

Example 59:

Change the order of integration $\int_0^a \int_{a-\sqrt{(a^2-y^2)}}^{a+\sqrt{(a^2-y^2)}} dy\, dx$.

Solution:

In the given integral the limits of ıntegration of x are given by $x = a - \sqrt{(a^2 - y^2)}$ and $x = a + \sqrt{(a^2 - y^2)}$ anzd those of y are given by $y = 0$ and $y = a$.

When $x = a - \sqrt{(a^2 - y^2)}$

or $x = a + \sqrt{(a^2 - y^2)}$,

we have $(x - a)^2 = a^2 - y^2$ or $(x - a)^2 + y^2 = a^2$

or $y^2 = 2ax - x^2$ which is a circle with center (a, 0) and radius a.

To reverse the order of integration, we divide the area OMNO into strips parallel to the y-axis. These strips will have their extremities on the x-axis and on the circular arc given by $y = \sqrt{(2ax - x^2)}$. Also x will go from 0 to 2a.

Hence, changing the order of integration, the given double integral transforms to $\int_0^{2a} \int_0^{\sqrt{(2ax-x^2)}} dx\, dy$.

Example 60:

Change the order of integration in the double integral $\int_0^\infty \int_x^\infty \frac{e^{-y}}{y} dx\, dy$ *and hence find its value.*

Solution:

In the given integral the limits of integration area given by the lines $y = x$, $y = \infty$, $x = 0$ and $x = \infty$. Therefore the region of integration is bounded by $x = 0$, $y = x$ and, an infinite boundary. In the given integral the limits of integration of y are variable while those of x are constant. Thus, we have to first integrate with respect to y regarding x as constant and then we integrate w.r.t. x. This is done by first integrating w.r.t. y along strip drawn parallel to the y-axis and then integrating w.r.t. x along all such strips so drawn as to cover the whole region of integration.

If we want to reverse the order of integration, we have the to first integrate w.r.t. x regarding y as constant and then we integrate w.r.t. y. This is done by dividing this area into strips parallel to the x-axis. So we take strips parallel to the x-axis. Starting from the line x = 0 and terminating on the line y = x. Now the limits for x are 0 to y and the limits for x are 0 to y and the limits for y are 0 to ∞.

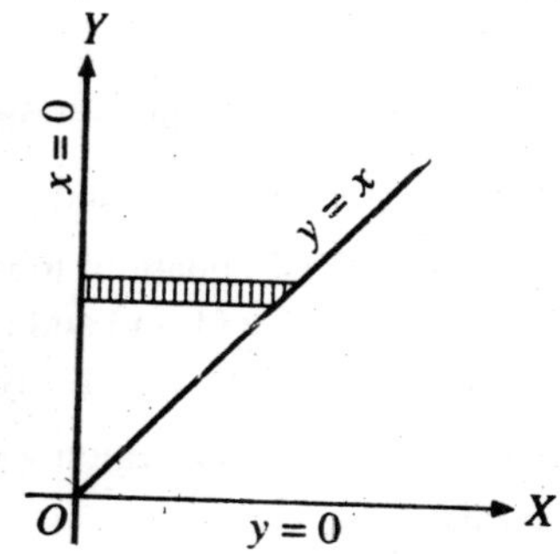

Fig. 1.11

Hence by changing the order of integration, we have

$$\int_0^\infty \int_x^\infty \frac{e^{-y}}{y} dx\,dy = \int_0^\infty \int_0^y \frac{e^{-y}}{y} dy\,dx$$

$$= \int_0^\infty \frac{e^{-y}}{y} [x]_0^y dy = \int_0^\infty \frac{e^{-y}}{y} \cdot y\,dy = \int_0^\infty e^{-y} dy = \left[\frac{e^{-y}}{-1}\right]_0^\infty = 1.$$

Example 61:

Transform $\int_0^a \int_0^{a-x} f(x, y) dx\,dy$, *by substitution* $x + y = u$, $y = uv$.

Solution:

We have $\iint f(x, y) dx\,dy = \iint F(u, v) u\,du\,dv$.

Now in the given integral, the region of integration is bounded by the lines y = 0, y = a – x, x = 0 and x = a.

Put x = u – y = u – uv = u (1 – v) and y = uv.

Then in the uv-plane the four straight lines become uv = 0, uv = a – u (1 – v), u (1 – v) = 0 and u (1 – v) = a, giving v = 0, v = 1, u = 0 and u = a.

Hence for the given region, v varies from 0 to 1 and u varies from 0 to a.

Therefore, by changing the variables, the given double integral transforms to $\int_0^a \int_0^1 F(u, v) u\,du\,dv$.

Example 62:

By using the transformation $x + y = u$, $y = uv$, *show that*

$$\int_0^1 \int_0^{1-x} e^{y/(x+y)} dx\,dy = \frac{1}{2}(e - 1).$$

Solution:

We have dx dy = u du dv.

Here the region of integration is bounded by the lines

y = 0, y = 1 – x, x = 0 and x = 1.

Changing these equations to new variables u and v by using the relations x = u – y = u – uv = u (1 – v) and y = uv, we have uv = = 0, uv = 1 – u (1 – v), u (1 – v) = 0 and u (1 – v) = 1, giving v = 0, = 1, u = 0 and u = 1.

Hence for the given region v varies from 0 to 1 and u varies from 0 to 1.

Further $e^y/(x + y) = e^{uv/u} = e^v$. $[\because x + y = u, y = uv]$

Therefore, changing the variables to u, v, the given integral becomes

$$= \int_0^1 \int_0^1 e^v . u\,du\,dv = \int_0^1 \left[e^v\right]_0^1 u\,du = \int_0^1 (e^1 - e^0) u\,du$$

$$= (e - 1) \int_0^1 u\,du = (e - 1). \left[\frac{u^2}{2}\right]_0^1 = \frac{1}{2}(e - 1)$$

Example 63:

Change the order of integration in the integral

$$\int_0^{a\cos\alpha} \int_{x\tan\alpha}^{\sqrt{(a^2 - x^2)}} f(x, y)\,dx\,dy.$$

Solution:

In the given integral the limits of integration of y are given by y = x tan α which is a straight line passing through the origin and

$$y = \sqrt{(a^2 - x^2)}$$

i.e., $y^2 = a^2 - x^2$

i.e., $x^2 + y^2 = a^2$

which is a circle of radius a with centre at the origin (0, 0).

Again the limits of integration of x are given by x = 0 *i.e.*, the y-axis and x = a cos α which is a straight line parallel to the y-axis at a distance a cos from the origin.

We draw the curve y = x tan α, $x^2 + y^2 = a^2$, x = 0 and x = a cos α, giving the limits of integration, in the same figure. We observe that the region of integration is the area OMNO.

In the given integral the limits of integration of y are variable while those of x are constant. Thus we have to first integrate with respect to y regarding x as constant and then we integrate w.r.t. x. This is done by covering the area of integration OMNO by drawing the straight lines x = constant *i.e.*, by dividing this area into strips parallel to the y-axis.

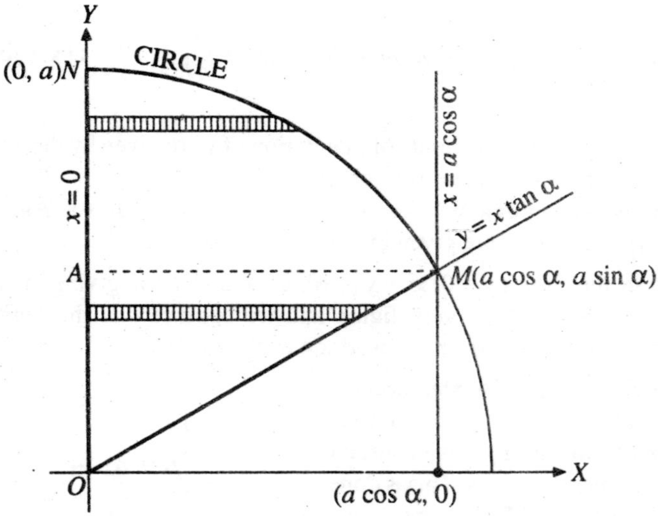

Fig. 1.12

If we want to reverse the order of integration, we have to first integrate with respect to x regarding yas constant and then we integrate w.r.t. y. This is done by covering the area of integration OMNO by drawing the straight lines y = constant *i.e.*, by dividing this area into strips parallel to the x-axis.

Now if we take strips parallel to the x-axis starting from the line x = 0; some of these strips end on the line OM while the others end on the arc MN of the circle $x^2 + y^2 = a^2$. So we draw the line of demarcation MA dividing the area OMNO into two portions OMA and AMN.

For the point M, x = a cos α. Putting x = a cos α in the equation of the line y = x tan α, we get y = a sin α. So the y-coordinate of the point M is a sin α and the equation of the line of demarcation MA is y = a sin α.

For the region OMA, x varies from 0 to y cot α and y varies from 0 to a sin α.

For the region AMN, x varies from 0 to $\sqrt{(a^2 - y^2)}$ and y varies from a sin α to a.

Therefore, changing the order of integration, the given double integral transforms to

$$\int_0^{a\sin\alpha}\int_0^{y\cot\alpha} f(x,y)\,dy\,dx + \int_{a\sin\alpha}^{a}\int_0^{\sqrt{(a^2-y^2)}} f(x,y)\,dy\,dx.$$

Example 64:

Change the order of integration in the integral $\int_0^a\int_0^{\sqrt{(a^2-x^2)}} f(x,y)\,dx\,dy.$

Solution:

In the given integral the limits of integration of y are given by the straight line $y = 0$ (*i.e.*, the x-axis) and the curve $y = \sqrt{(a^2 - x^2)}$ *i.e.*, $y^2 = a^2 - x^2$ *i.e.*, $x^2 + y^2 = a^2$ which is circle with centre at the origin and radius a. Again the limits of integration of x are given by the lines $x = 0$ and $x = a$.

We draw the curves $y = 0$, $x^2 + y^2 = a^2$, $x = 0$ and $x = a$, giving the limits of integration, in the same very figure and we observe that the region of integration is the area OAB of the quadrant of the circle $x^2 + y^2 = a^2$.

To change the order of integration in the given integral, we have to first integrate w.r.t. x regarding y as a constant and then we integrate w.r.t. y. This is done by covring the area OAB by strips drawn parallel to the x-axis. These strips start from the line OB (*i.e.*, $x = 0$) and terminate on the arc AB of the circle $x^2 + y^2 = a^2$. So on these strips x varies from 0 to $\sqrt{(a^2 - y^2)}$. Also to cover the area OAB, y varies from 0 to a. Hence by changing the order of integration, we have the given integral

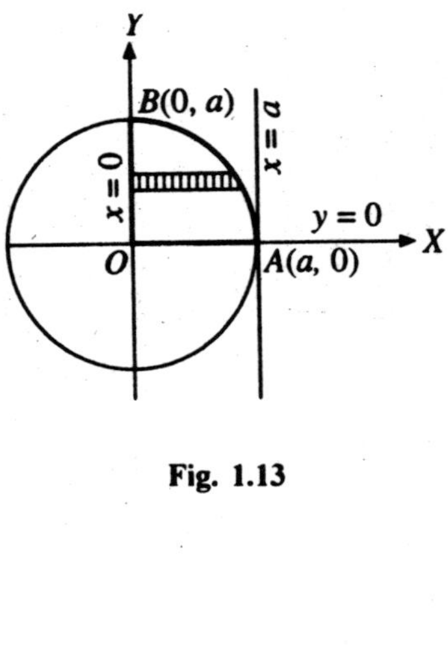

Fig. 1.13

$$= \int_0^a\int_0^{\sqrt{(a^2-x^2)}} f(x,y)\,dx\,dy.$$

Example 65:

Change the order of integration in $\int_0^a\int_{mx}^{lx} f(x,y)\,dx\,dy.$

Solution:

Here the area of integration is bounded by the straight lines $y = mx$, $y = lx$, $x = 0$ and $x = a$. Drawing all these lines in one figure, we observe that area of integration is OABO.

To reverse the order of integration, cover this area OABO by strips parallel to the axis of x. Draw the straight line AN parallel to the x-axis and thus divide the area OABO into two portions OAN and NBA according to the character of the strips.

For the point A, $x = a$.

Putting $x = a$ in the equation of the line $y = mx$,

we get $y = ma$.

Also for the point B, $x = a$;

therefore putting $x = a$ in the equation of the line $y = lx$, we get $y = la$.

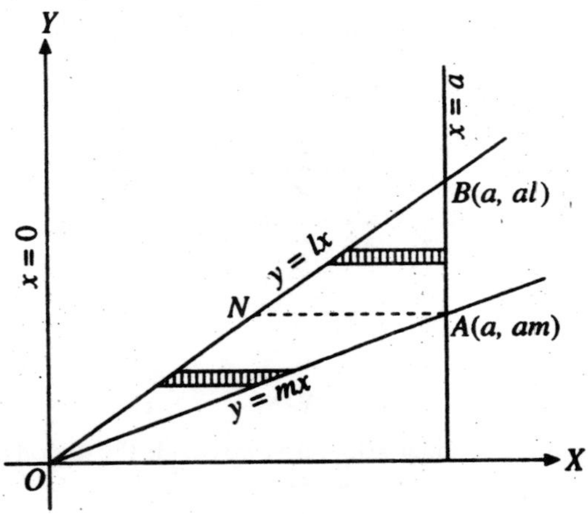

Fig. 1.14

Now for the area ONA, x varies from the line $y = lx$ to $y = mx$ *i.e.*, x varies from y/l to y/m and y varies from 0 to am. Again for the area NBA, x varies from the line NB ($y = lx$) to the line $x = a$ *i.e.*, x varies from y/l to a and y varies from am to al.

Therefore, by changing the order of integration the given integral transforms to

$$\int_0^{am}\int_{y/l}^{y/m} f(x,y)\,dy\,dx + \int_{am}^{al}\int_{y/l}^{a} f(x,y)\,dy\,dx.$$

Example 66:

Change the order of integration in $\int_0^{2a}\int_{x^2/4a}^{3a-x} f(x,y)\,dx\,dy$.

Solution:

In the given integral the limits of integration are given by $x^2/4a = y$ *i.e.*, $x^2 = 4ay$, (which is a parabola passing through the origin), and the lines

$y = 3a - x$, $x = 0$, and $x = 2a$. Drawing these curves in one figure we observe that the region of integration is the area OABMO.

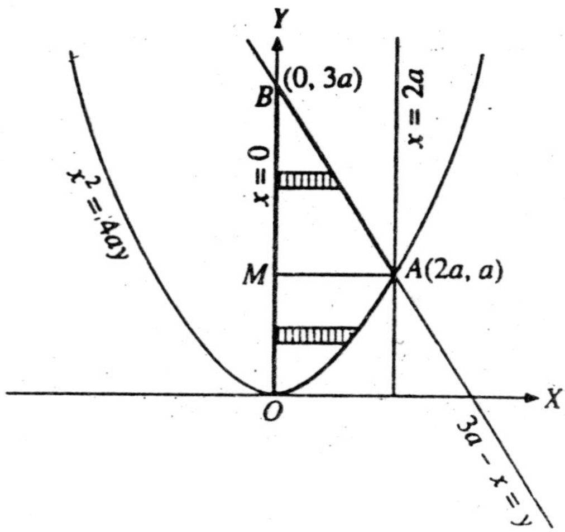

Fig. 1.15

To change the order of integration, first we divide the region of integration into two portions OAM and MAB, by drawing the line AM parallel to the x-axis. Now to reverse the order of integration, cover the whole region OABMO by strips parallel to the x-axis starting from the line $x = 0$. Some of these strips end on the arc OA while others end on the line AB.

For the point A, we have $x = 2a$. Putting $x = 2a$ in the equation of the line $y = 3a - x$, we get $y = a$.

For the region OAM , x varies form 0 to $\sqrt{(4ay)}$ and y varies form 0 to a. Again for the region MAB, x varies form 0 to $3a - y$ and y varies from a to 3a.

Hence the transformed integral is given by

$$\int_0^a \int_0^{\sqrt{(4ay)}} f(x,y) dy\, dx + \int_a^{3a} \int_0^{3a-y} f(x,y) dy\, dx.$$

Example 67:

Change the order of integration $\int_0^a \int_{x^2/a}^{2a-x} xy\, dx\, dy$.

Solution:

In the given integral the limits of integration are given by $x^2/a = y$ *i.e.*, $x^2 = ay$ (which is a parabola passing through the origin), and the straight lines $y = 2a - x$, $x = 0$ and $x = a$.

Here the coordinates of A are (a, a) and those of B are (0, 2a).

The transformed integral is given by

$$\int_0^a \int_0^{\sqrt{(ay)}} xy\,dy\,dx + \int_a^{2a} \int_0^{2a-y} xy\,dy\,dx.$$

Example 68:

Change the order of integration in the double integral

$$\int_0^a \int_0^{b/(b+x)} f(x,y)dx\,dy.$$

Solution:

In the given integral the limits of integration of y are given by y = 0 (*i.e.*, the x-axis) and y = b/(b + x) *i.e.*, y (b + x) = b which is a rectangular hyperbola having for its asymptotes the straight lines y = 0 and x = – b. Again the limits of integration of x are given by the straight lines x = 0 (*i.e.*, the y-axis) and x = a. We draw the curves y (b + x) = b, y = 0, x = 0 and x = a, giving the limits of integration, in the same figure. We observe that the region of integration is the area OMABO.

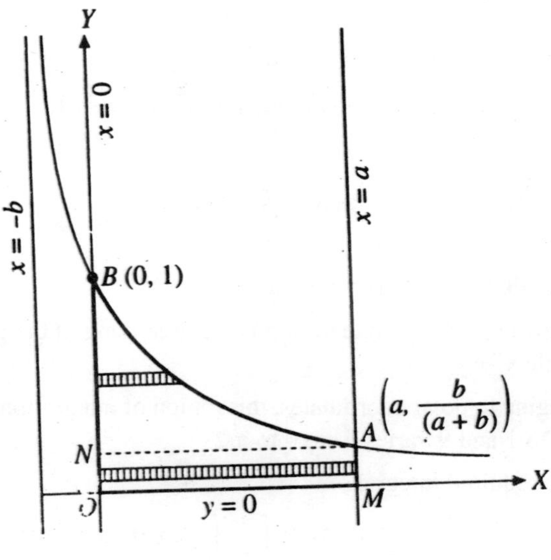

Fig. 1.16

In the given integral we are required to integrate first w.r.t. y and then w.r.t x. To change the order ofintegration, we have to first integrate w.r.t.

regarding y as constant and then we integrate w.r.t. y. This is done by covering the area of integration OMABO by drawing the straight lines y = constant *i.e.*, by dividing this area into strips parallel to the x-axis.

Now, if we take strips parallel to the x-axis originating from the line x = 0, some of these strips terminate on the line AM while the others terminate on the arc AB. So according to the character of the strips we divide the region of integration into two portions namely NOMA and NAB, by drawing the line AN parallel to the axis of x.

For the point B, x = 0. Putting x = 0 in the equation y (b + x) = b, we get y = 1. So the coordinates of the point B are (1, 1).

Similarly, putting x = a in the equation y (b + x) = b, we get y = b/(a + b) and thus the coordinates of the point Aare (a, b/(a + b)).

For the area NBA, x varies form 0 to b (1 – y)/y and y varies from b/(a + b) to 1.

Therefore, changing the order of integration, the given double integral transforms to

$$\int_0^{b/(a+b)}\int_0^a f(x,y)dy\,dx + \int_{b/(a+b)}^1\int_0^{b(!-y)/y} f(x,y)dy\,dx.$$

Example 69:

Evaluate $\iint xy(x^2+y^2)^{3/2}\,dx\,dy$

over the positive quadrant of the circle $x^2 + y^2 = 1$.

Solution:

Changing to polers by putting x = r cos θ,

y = r sin θ, we have J = r

so that dxdy = J dθ dr = r dθ dr.

The given region of integration is the area lying in the positive quadrant of the circle $x^2 + y^2 = 1$.

Changing to polar coordinates, this region of integration is covered when r varies 0 to 1 and θ varies from 0 to π/2.

∴ the required integral

$$\iint xy(x^2+y^2)^{3/2}\,dx\,dy = \int_0^{\pi/2}\int_0^1 r\cos\theta.r\sin\theta.(r^2)^{3/2}.r\,d\theta\,dr$$

$$= \int_0^{\pi/2}\int_0^1 r^6\sin\theta\cos\theta\,d\theta\,dr = \int_0^{\pi/2}\left[\frac{r^7}{7}\right]_0^1 \sin\theta\cos d\theta$$

$$=\frac{1}{7}\int_0^{\pi/2}\frac{1}{2}\sin 2\theta d\theta=\frac{1}{14}\left[-\frac{\cos 2\theta}{2}\right]_0^{\pi/2}=-\frac{1}{28}[-1-1]=\frac{1}{14}.$$

Example 70:

Evaluate $\iint\sqrt{(a^2-x^2-y^2)}dx\,dy$

over the semi-circle $x^2+y^2=ax$ in the positive quadrant.

Solution:

Here the region of integration is a semi-circle. Therefore, for the sake of convenience, changing to polar by putting $x = r\cos\theta$ and $y = r\sin\theta$ in $x^2 + y^2 = ax$, we have $r^2\cos^2\theta + r^2\sin^2\theta = a\,r\cos\theta$ or $r^2(\sin^2\theta + \cos^2\theta) = a\,r\cos\theta$

or $\quad r = a\cos\theta.$

The equation $r = a\cos\theta$ represents a circle passing through the pole and diameter through the pole along the initial line.

For the given region r varies from 0 to $a\cos\theta$ and θ varies from 0 to $\pi/2$.

$$\therefore \iint\sqrt{(a^2-x^2-y^2)}dx\,dy=\int_0^{\pi/2}\int_0^{a\cos\theta}\sqrt{(a^2-r^2)}.r\,d\theta\,dr,\quad [\because x^2+y^2=r^2 \text{ and } dx\,dy=r\,d\theta\,dr]$$

$$=\int_0^{\pi/2}\left[\int_0^{a\cos\theta}-\frac{1}{2}(a^2-r^2)^{1/2}.(-2r)dr\right]d\theta$$

$$=\int_0^{\pi/2}\left[-\frac{1}{2}\cdot\frac{2}{3}(a^2-r^2)^{3/2}\right]_0^{a\cos\theta}d\theta.$$

$$=-\frac{1}{3}\int_0^{\pi/2}(a^2\sin^3\theta-a^3)d\theta=-\frac{a^3}{3}\left[\frac{2}{3.1}-\frac{\pi}{2}\right]$$

$$=\frac{1}{3}a^3\left(\frac{1}{3}\pi-\frac{2}{3}\right).$$

Example 71:

Changed the order of integration in $\int_0^a\int_x^{a^2/x}f(x,y)dx\,dy.$

Solution:

In the given integral the limits of integration of y are given by y = x which is a straight line passing through the origin equally inclined to both the axes and $y = a^2/x$ or $xy = a^2$ which is a rectangular hyperbola. Again the limits of integration of x are given by the straight lines x = 0 (*i.e.*, the y-axis) and x = a.

We draw the curves $y = x$, $xy = a^2$, $x = 0$, $x = 0$, and $x = a$, giving the limits of integration, in the same figure. We observe that the region of integration is the area LMOY... extended upto infinity on the above side.

In the given integral we are required to integrate first w.r.t.; y and then w.r.t. x. If we want to change the order of integration, we have to first integrate w.r.t. x regarding y as constant and then we integrate w.r.t. y This is done by covering the area of integration by strips parallel to the x-axis.

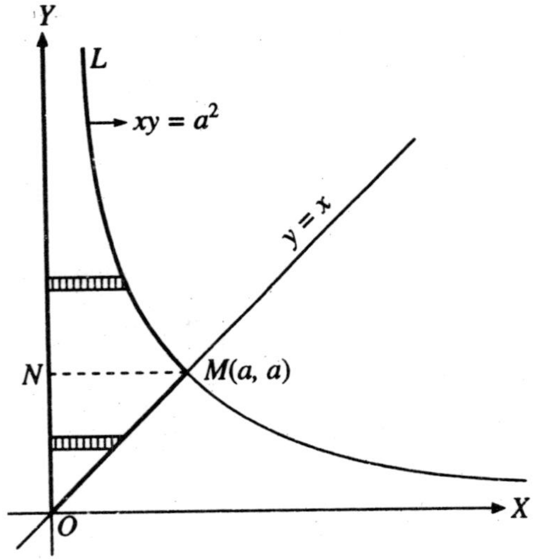

Fig. 1.17

Now if we take strips parallel to the x-axis starting from the line $x = 0$,some of these strips end on the in OM while the others end on the arc ML of the rectangular hyperbola. So we divide the region of integration into two portions, the triangle OMN and the area YNML which extends upto infinity, by drawing the line MN parallel to the axis of x.

For the point M, $x = a$. Putting $x = a$ in the equation of the line $y = x$ or the rectangular hyperbola $xy = a^2$, we get $y = a$.

So the y-coordinate of the point M is a and the equation of the line of demarcation MN is $y = a$.

For the area OMN, x varies from 0 to y and y varies from 0 to a.

For the area YMNL..., x varies from 0 to a^2/y and y varies from a to ∞.

Hence by changing the order of integration, we have the given integral

$$= \int_0^a \int_0^y f(x,y)\,dy\,dx + \int_a^\infty \int_0^{a^2/y} f(x,y)\,dy\,dx.$$

Example 72:

Change the order of integration in

$$\int_0^a \int_{(b/a)\sqrt{(a^2-x^2)}}^b f(x,y)\,dx\,dy, \text{ where } c < a.$$

Solution:

In the given integral the limits of integration of y are given by

$$y = \frac{b}{a}\sqrt{(a^2 - x^2)}\ i.e.,\ \frac{x^2}{a^2} + \frac{y^2}{b^2} = 1$$

which is an ellipse with centre (0, 0) and the straight line y = b.

Again the limits of integration of x are given by the straight lines x = c and x = a.

Draw the ellipse $x^2/a^2 + y^2/y^2 = 1$ and the straight lines y = b, x = c and x = a, bounding the region of integration, in the same figure. We observe that the region of integration is the area ABECA. In the given integral, the limits of integration of y being variable, we are required to integrate first w.r.t. y and then w.r.t. x.

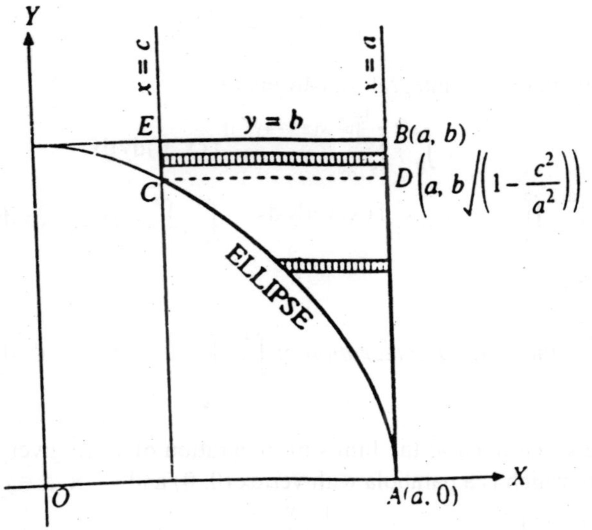

Fig. 1.18

In order to integrate in the reverse order, divide whole the area into strips parallel to the x-axis originating either form the EC (*i.e.*, x = c) or form the arc

AC of the ellipse and terminating on the line BA (*i.e.*, x = a). While integrating we must first obviously divide the region of integration ABEA into two portions AD and ECDB according to the character of the strips. For the point C, x = c. Putting x = c in the equation of the ellipse $x^2/a^2 + y^2/b^2 = 1$, we get $y = b\sqrt{\{1 - (c^2/a^2)\}}$ which is the y-coordinate of the point C. The equation of the line of demarcation CD is thus $y = b\sqrt{\{1 - (c^2/a^2)\}}$.

For the area CAD, x varies from $a\sqrt{\{1 - (y^2/b^2)\}}$ to a and y varies from 0 to $b\sqrt{\{1 - (c^2/a^2)\}}$.

For the area ECDB, x varies from c to a and y varies form $b\sqrt{\{1 - (c^2/a^2)\}}$ to b.

Therefore, changing the order of integration, the given double integral transforms to

$$\int_0^{b\sqrt{\{1-(c^2/a^2)\}}} \int_{a\sqrt{\{1-(y^2/b^2)\}}}^{a} f(x,y)\,dy\,dx + \int_{b\sqrt{\{1-(c^2/a^2)\}}}^{a} \int_c^a f(x,y)\,dy\,dx.$$

Example 73:

Change the order of integration in the double integral

$$\int_0^{2a} \int_{\sqrt{(2ax-x^2)}}^{\sqrt{(2ax)}} f(x,y)\,dx\,dy.$$

Solution:

The given double integral transforms to

$$\int_0^{a/2} \int_{y^2/a}^{\frac{1}{2}[a-\sqrt{(a^2-4y^2)}]} f(x,y)\,dy\,dx$$

$$= \int_0^{a/2} \int_{\frac{1}{2}[a+\sqrt{(a^2-4y^2)}]}^{a} f(x,y)\,dy\,dx + \int_{a/2}^{a} \int_{y^2/a}^{a} f(x,y)\,dy\,dx.$$

Example 74:

Change the order of integration in $\int_0^{a/2} \int_{x^2/a}^{x-(x^2/a)} f(x,y)\,dx\,dy.$

Solution:

In the given integral the limits of integration of y are given by $y = x^2/a$ *i.e.*, $x^2 = ay$ which is a parabola with vertex (0, 0) and $x - x^2/a = y$ *i.e.*, $ax - x^2 = ay$ *i.e.*, $\left(x - \frac{1}{2}a\right)^2 = -a\left(y - \frac{1}{4}a\right)$ which is also a parabola with vertex $\left(\frac{1}{2}a, \frac{1}{4}a\right)$.

The points of intercession of the two parabolas are (0, 0) and $\left(\frac{1}{2}a, \frac{1}{4}a\right)$.

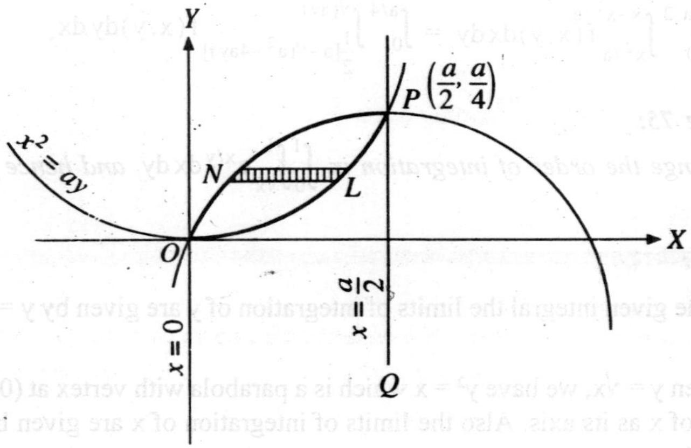

Fig. 1.19

Again the limits of integration of x are given by x = 0 *i.e.*, the y-axis and x = a/2 which is a straight line parallel to the y-axis at a distance a/2 from the origin.

Draw the two parabolas $x^2 = ay$ and $\left(x - \frac{1}{2}a\right)^2 = -a\left(y - \frac{1}{4}a\right)$ intersecting at O (0, 0) and P $\left(\frac{1}{2}a, \frac{1}{4}a\right)$ along with the lines x = 0 and x = a/2 in the same figure. We observe that the region of integration is ONPLO. In the given integral we are required to integrate first w.r.t. y ($\because$ the limits of integration of y are variable) and then w.r.t. x. To reverse the order of integration, draw strips parallel to the x-axis originating from the arc ONP of the parabola $ax - x^2 = ay$ and terminating on the arc OLP of the parables $x^2 = ay$. Then for the region ONPLO, the limits of integration for x are given by $ax - x^2 = ay$ and $x^2 = ay$. Solving $ay = ax - x^2$ *i.e.*, $x^2 - ax + ay = 0$ for x, we get

$$x = \frac{1}{2}\left[a \pm \sqrt{(a^2 - 4ay)}\right]$$

or $$x = \frac{1}{2}\left[a - \sqrt{(a^2 - 4ay)}\right],$$

rejecting the +ive sign since x cannot be greater than $\frac{1}{2}$ a in the region ONPLO.

Thus the limits of x are $x = \frac{1}{2}\left[a - \sqrt{(a^2 - 4ay)}\right]$ and $x = \sqrt{(ay)}$.

Clearly for this region y varies from 0 to $\frac{1}{4}a$.

Hence by changing the order of integration, we have

$$\int_0^{a/2}\int_{x^2/a}^{x-x^2/a} f(x,y)\,dx\,dy = \int_0^{a/4}\int_{\frac{1}{2}[a-\sqrt{(a^2-4ay)}]}^{\sqrt{(ay)}} f(x,y)\,dy\,dx.$$

Example 75:

Change the order of integration in $\int_0^1\int_{\sqrt{x}}^1 e^{x/y}\,dx\,dy$ *and hence find its value.*

Solution:

In the given integral the limits of integration of y are given by $y = \sqrt{x}$ and $y = 1$.

When $y = \sqrt{x}$, we have $y^2 = x$ which is a parabola with vertex at (0, 0) and the axis of x as its axis. Also the limits of integration of x are given by $x = 0$ and $x = 1$

The region of integration is the area OABO.

To reverse the order of integration, we divide the area OABO into strips parallel to the x-axis.

Changing the order of integration, the given double integral I transforms to

$$I = \int_{y=0}^{1}\int_{x=0}^{y^2} e^{x/y}\,dy\,dx.$$

We have

$$I = \int_{y=0}^{1}\left[y e^{x/y}\right]_{x=0}^{y^2} dy$$

$$= \int_0^1 (ye^y - y)\,dy$$

$$= \left[ye^y - e^y - \frac{y^2}{2}\right]_0^1$$

$$= e - e - \frac{1}{2} + 1 = \frac{1}{2}.$$

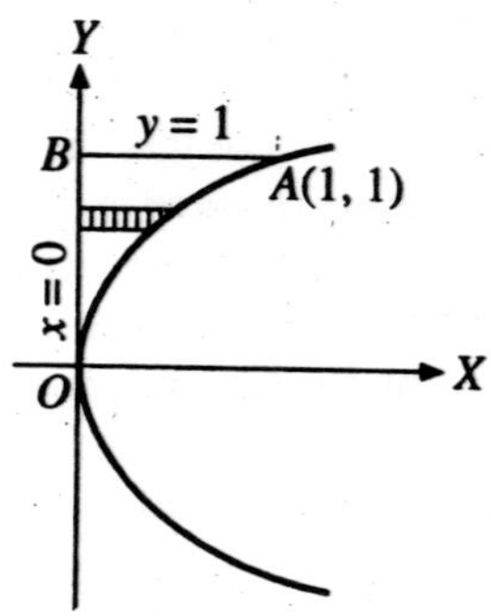

Fig. 1.20

Example 76:

Change the order of integration $\int_0^a\int_{x/a}^{\sqrt{(x/a)}} (x^2 + y^2)\,dx\,dy.$

Solution:

In the given integral the limits of integration of y are given by y = x/a and y = √(x/a). y = x/a is a straight line passing through the origin.

When y = √(x/a), we have $y^2 = x/a$ which is a parabola with vertex at (0, 0) and x-axis as its axis.

The straight line y = x/a meets $y^2 = x/a$ at the points A(a, 1).

The limits of integration of x are given by x = 0 and x = a.

Thus, the region of integration is the area OABO.

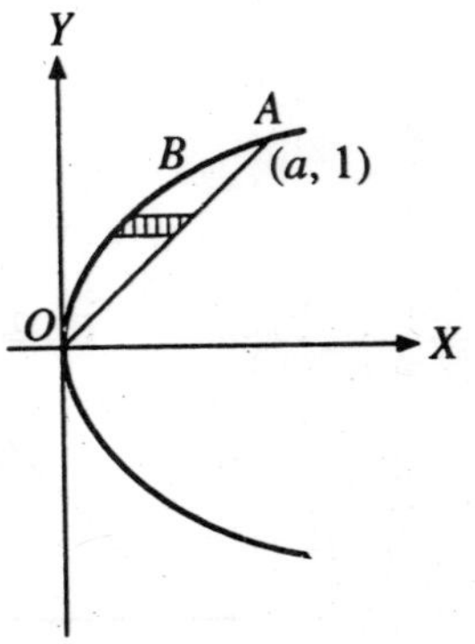

Fig. 1.21

To reverse the order of integration, we divide the area OABO into strips parallel to x-axis.

Changing the ordure of integration, the given double integral I transforms to

$$I = \int_{y=0}^{1} \int_{x=ay^2}^{ay} (x^2 + y^2)\,dy\,dx.$$

Example 77:

Change the order of integration in $\int_0^a \int_{\sqrt{(a^2-x^2)}}^{x+2a} f(x,y)\,dx\,dy.$

Solution:

Here the area of integration is bounded by the curves $y = \sqrt{(a^2 - x^2)}$

i.e., $x^2 + y^2 = a^2$

which is a circle with centre (0, 0) and radius a, y = x + 2a which is a straight line passing through (0, 2a), x = 0 *i.e.*, the y-axis and the line x = a which a line parallels to the y-axis at a distance a from the origin.

We draw the curves $x^2 + y^2 = a^2$, y = x + 2a, x = 0 and x = a, giving the limits of integration, in the same figure. We observe that the region of integration is the area MLANM.

To reverse the order of integration, cover this area of integration MLANM by strips parallel to the x-axis. Draw the lines MC and NB parallels to the x-axis so that the region of integration MLANM is divided into three portions MLC, NMCB and NAB.

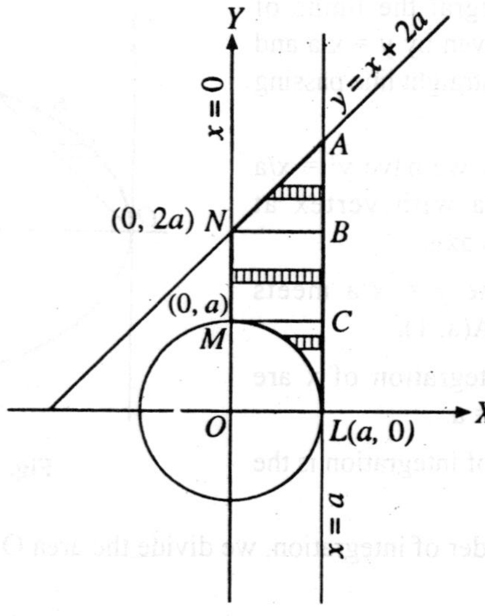

Fig. 1.22

For the region MLC, x varies from the arc ML of the circle $x^2 + y^2 = a^2$ to the line x = a *i.e.*, x varies from $\sqrt{(a^2 - y^2)}$ to a and y varies from 0 to a.

For the region NMCB, x varies from 0 to a and y varies from a to 2a.

For the region NBA, x varies from y – 2a to a and y varies from 2a to 3a.

Therefore, changing the order of integration, the given integral transforms to

$$\int_0^a \int_{\sqrt{(a^2-x^2)}}^a f(x,y)dx\,dy + \int_a^{2a}\int_0^a f(x,y)dy\,dx + \int_{2a}^{3a}\int_{y-2a}^a f(x,y)dy\,dx.$$

Example 78:

Change the order of integration in the double integral

$$\int_0^{ab/\sqrt{(a^2+b^2)}} \int_0^{(a/b)\sqrt{(b^2-y^2)}} f(x,y)dy\,dx.$$

Solution:

In the given integral the limits of integration of x are given by x = 0 *i.e.*, the y-axis and $x = (a/b)\sqrt{(b^2 - y^2)}$ *i.e.*, $x^2/a^2 + y^2/b^2 = 1$ which is an ellipse with centre as origin.

Again the limits ofintegration of y are given by y = 0 *i.e.*, the x-axis and $y = ab/\sqrt{(a^2 + b^2)}$ which is a straight line parallel to the x-axis at a distance $ab/\sqrt{(a^2 + b^2)}$ from the origin.

We draw the curves $x = 0$, $x^2/a^2 + y^2/b^2 = 1$, $y = 0$

and $y = ab/\sqrt{(a^2 + b^2)}$, giving the limits of integration, in the same figure. We observe that the region of integration is the area OPBAO.

In the given integral the limits of integration of x are variable while those of y are constant. Thus we have to first integrate w.r.t x regarding y as a constant and then we integrate w.r.t. y.

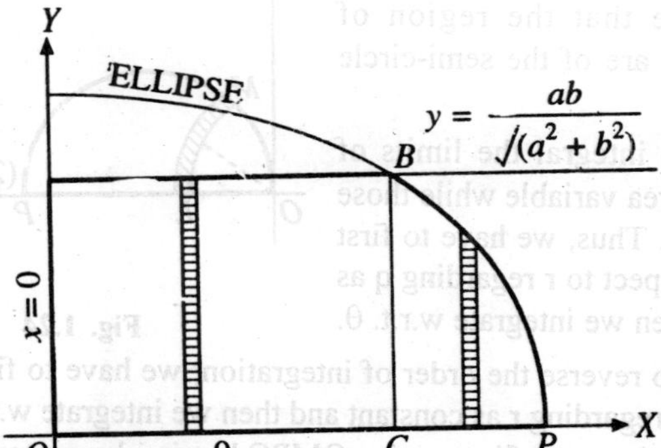

Fig. 1.23

If we want to reveres the order of integration, we have to first integrate w.r.t. y regarding x as constant and then we integrate w.r.t. x. This is done by covering the area of integration OPBAO by strips parallel to the y-axis. Now if we take strips parallel to the y-axis starting from the line $y = 0$, some of these strips and on the line AB while the others end on the arc BP of the ellipse. So we draw the line of demarcation BC dividing the area OPBAO into two portions OCBA and BCP. For the point B, $y = ab/\sqrt{(a^2 + b^2)}$. Putting this value of y in the equation of the ellipse $x^2/a^2 + y^2/b^2 = 1$, we get $x = ab/\sqrt{(a^2+b^2)}$. For the region OCBA, y varies from 0 to $(ab)/\sqrt{(a^2 + b^2)}$ and x varies from 0 to $ab/\sqrt{(a^2 + b^2)}$ to a.

Hence the given integral transforms to

$$\int_0^{ab/\sqrt{(a^2+b^2)}} \int_0^{ab/\sqrt{(a^2+b^2)}} f(x,y)\,dy\,dx$$

$$+ \int_{ab/\sqrt{(a^2+b^2)}}^{0} \int_a^{(b/a)\sqrt{(a^2-x^2)}} f(x,y)\,dx\,dy.$$

Example 79:

Change the order of integration in $\int_0^{\pi/2} \int_0^{2a\cos\theta} f(r,\theta)\,d\theta\,dr.$

Solution:

Here the region of integration is bounded by the polar curve $r = 0$ (the pole), $r = 2a \cos \theta$ (a circle of diameter 2a passing through the pole), $\theta = 0$ (the initial line) and $\theta = \pi/2$ (a line through the pole perpendicular to initial line).

We draw the curves $r = 0$, $r = 2a \cos \theta$, $\theta = 0$ and $\theta = \pi/2$, giving the limits of integration, in the same figure.

We observe that the region of integration is the are of the semi-circle OMPO.

In the given integral the limits of integration of r area variable while those of θ are constant. Thus, we have to first integrate with respect to r regarding q as a constant and then we integrate w.r.t. θ.

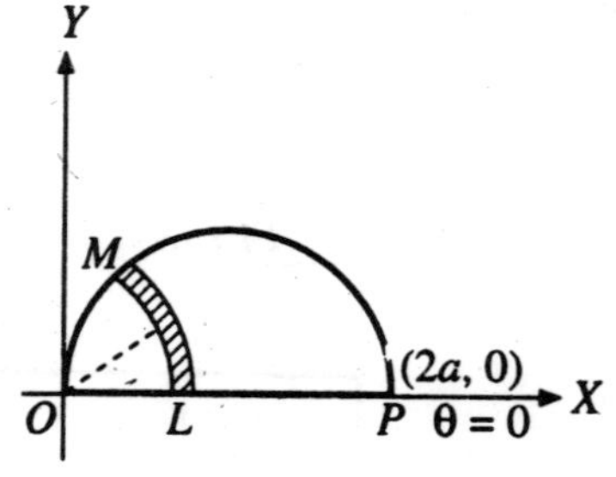

Fig. 1.24

If we want to reverse the order of integration, we have to first integrate with respect to θ regarding r as constant and then we integrate w.r.t. r. This is done by covering the area of integration OMPO by circular arcs with centre as pole. On these arcs θ varies and r remains constant. Thus, for the area OMPO, for a fixed value of r, θ varies from the initial line (*i.e.*, $\theta = 0$ to a point on the arc OMP of the circle $r = 2a \cos \theta$ *i.e.*, to a point for which $\theta = \cos^{-1}(r/2a)$ and r varies from 0 to 2a.

Hence by chaining the order of integration, we have

$$\int_0^{\pi/2} \int_0^{2a\cos\theta} f(r,\theta)\, d\theta\, dr = \int_0^{2a} \int_0^{\cos^{-1}(r/2a)} f(r,\theta)\, dr\, d\theta.$$

Example 80:

Transform $\iint f(x,y)\,dx\,dy$ by the substitution $x + y = u$, $y = uv$.

Solution:

We have $x + y = u$ and $y = uv$. ...(1)

From these, we have $x = u - y = u - uv$

and $y = uv$. ...(2)

$$\therefore \frac{\partial x}{\partial u} = 1 - v,$$

$$\frac{\partial x}{\partial v} = -u, \frac{\partial y}{\partial u} = v$$

and $\quad \dfrac{\partial y}{\partial v} = u.$

$$\therefore J = \frac{\partial(x,y)}{\partial(u,v)} = \begin{vmatrix} \dfrac{\partial x}{\partial u} & \dfrac{\partial x}{\partial v} \\ \dfrac{\partial y}{\partial u} & \dfrac{\partial y}{\partial v} \end{vmatrix} = \begin{vmatrix} 1-v & -u \\ v & u \end{vmatrix} = u.$$

$\therefore$ dx dy = J du dv = u du dv.

Hence the given integral transforms to

$$\iint F(u,v)u\,du\,dv.$$

Example 81:

Transform $\iint f(x,y)dx\,dy$ *to polar coordinates.*

Solution:

We have x = r cos θ, y = r sin θ.

Now $\quad J = \dfrac{\partial(x,y)}{\partial(r,\theta)} = \begin{vmatrix} \dfrac{\partial x}{\partial r} & \dfrac{\partial x}{\partial \theta} \\ \dfrac{\partial y}{\partial r} & \dfrac{\partial y}{\partial \theta} \end{vmatrix} = \begin{vmatrix} \cos\theta & -r\sin\theta \\ \sin\theta & r\cos\theta \end{vmatrix} = r.$

$\therefore$ dx dy = Jdθ dr = r dθ dr.

Hence the given integral transforms to $\iint F(r,\theta)r\,d\theta\,dr$.

Example 82:

By using transformation x + y = u, y = uv, prove that

$$\iint \{xy(1-x-y)\}^{1/2}\,dx\,dy$$

taken over the area of the triangle bounded by the lines x = 0, y = 0, x + y = 1 is 2π/105.

Solution:

We have dx dy = ududv; u varies from 0 to 1 and also v varies from 0 to 1.

Now $\{xy(1-x-y)\}^{1/2} = [xy\{1-(x+y)\}]^{1/2}$

$= [u(1-v).uv.(1-u)]^{1/2}$ $\quad$ [$\because$ x = u (1 – v), y = uv]

$= u(1-u)^{1/2}.\,v^{1/2}(1-v)^{1/2}.$

Hence the given double integral transforms to

$$\int_0^1\int_0^1 u(1-u)^{1/2}.v^{1/2}(1-v)^{1/2}.u\,du\,dv$$

$$= \left[\int_0^1 u^2 (1-u)^{1/2}\, du\right] \cdot \left[\int_0^1 v^{1/2} (1-v)^{1/2}\, dv\right]$$

$$= \left[\int_0^1 u^{3-1} (1-u)^{3/2-1}\, du\right] \cdot \left[\int_0^1 v^{3/2-1} (1-v)^{3/2-1}\, dv\right]$$

$$= B\left(3, \frac{3}{2}\right) . B\left(\frac{3}{2}, \frac{3}{2}\right), \text{ [by the def. of Beta function]}$$

$$= \frac{\Gamma(3)\Gamma\left(\frac{3}{2}\right)}{\Gamma\left(3+\frac{3}{2}\right)} \cdot \frac{\Gamma\frac{3}{2}.\Gamma\frac{3}{2}}{\Gamma\left(\frac{3}{2}+\frac{3}{2}\right)} = \frac{2\left[\frac{1}{2}\sqrt{\pi}\right]^3}{\frac{7}{2}\cdot\frac{5}{2}\cdot\frac{3}{2}\cdot\frac{1}{2}\cdot\sqrt{\pi}.2} = \frac{2\pi}{105}.$$

Example 83:

Evaluate $\iint (x^2+y^2)^{7/2}\, dx\, dy$ *over the circle* $x^2 + y^2 = 1$.

Solution:

Here the region of integration is a circle. Therefore we shall change the given double integral to polar coordinates by putting $x = r \cos \theta$ and $y = r \sin \theta$. We have

$$J = \frac{\partial(x,y)}{\partial(r,\theta)} = \begin{vmatrix} \frac{\partial x}{\partial r} & \frac{\partial x}{\partial \theta} \\ \frac{\partial y}{\partial r} & \frac{\partial y}{\partial \theta} \end{vmatrix} = \begin{vmatrix} \cos\theta & r\sin\theta \\ \sin\theta & r\cos\theta \end{vmatrix} = r.$$

$\therefore$ dx dy = J dθ dr = r dθ dr.

Clearly, the region of integration is the circle $x^2 + y^2 = 1$ *i.e.*, the circle with centre (0, 0) and radius 1.

Changing to polar coordinates, the region of integration is covered when r varies from 0 to 1 and θ varies from 0 to 2π.

$$\therefore \iint_{x^2+y^2\le 1} (x^2+y^2)^{7/2}\, dx\, dy = \int_0^{2\pi}\int_0^1 (r^2)^{7/2} J\, d\theta\, dr$$

$$= \int_0^{2\pi}\int_0^1 r^7 . r . d\theta\, dr = \int_0^{2\pi}\int_0^1 r^8 d\theta\, dr = \int_0^{2\pi}\left[\frac{r^9}{9}\right]_0^1 d\theta$$

$$= \frac{1}{9}\int_0^{2\pi} d\theta = \frac{1}{9}[\theta]_0^{2\pi} = \frac{2}{9}\pi.$$

Example 84:

Evaluated $\iint e^{-(x^2+y^2)} dx\, dy$ *over the circle* $x^2 + y^2 = a^2$.

Solution:

Changing to polar coordinates, the equation $x^2 + y^2 = a^2$ transforms to $r^2\cos^2\theta + r^2\sin^2\theta = a^2$ *i.e.*, $r = a$.

Hence for the given region r varies from 0 to a and θ varies from 0 to 2π.

Also $\quad dx\,dy = r\,d\theta\,dr$.

$\therefore$ the required integral

$$\iint e^{-(x^2+y^2)}dx\,dy = \int_0^{2\pi}\int_0^a e^{-r^2}r\,d\theta\,dr$$

$$= \int_0^{2\pi}\int_0^{a^2} e^{-t}.\frac{1}{2}d\theta\,dt,$$

putting $r^2 = t$ so that $2r\,dr = dt$

$$= \frac{1}{2}\int_0^{2\pi}\left[\frac{e^{-t}}{-1}\right]_0^{a^2} d\theta = -\frac{1}{2}\int_0^{2\pi}(e^{-a^2} - 1)d\theta$$

$$= -\frac{1}{2}(e^{-a^2} - 1)[\theta]_0^{2\pi} = \frac{1}{2}(1 - e^{-a^2}).2\pi = \pi(1 - e^{-a^2}).$$

Example 85:

Evaluate the following double integrals:

(i) $\int_0^a\int_0^b (x^2 + y^2)dx\,dy$;

(ii) $\int_1^a\int_1^b \frac{dx\,dy}{xy}$;

(iii) $\int_1^2\int_0^x \frac{dx\,dy}{x^2 + y^2}$;

(iv) $\int_0^{\pi/2}\int_{\pi/2}^{\pi} \cos(x + y)dy\,dx$;

(v) $\int_0^1\int_0^{x^2} e^{y/x}dx\,dy$;

(vi) $\int_1^2\int_0^{3y} y\,dy\,dx$.

(vii) $\int_0^2\int_0^{2x-4} \frac{2y - 1}{x + 1}dx\,dy$

Solution:

(i) We have

$$\int_0^a\int_0^b(x^2+y^2)dx\,dy = \int_0^a\left[x^2y+\frac{y^3}{3}\right]_{y=0}^{b} dx,$$

(integration w.r.t. y treating x as constant)

$$= \int_0^a\left[bx^2+\frac{b^3}{3}\right]dx = \left[b\frac{x^3}{3}+\frac{b^3}{3}x\right]_0^a = \frac{ba^3}{3}+\frac{b^3a}{3}$$

$$= \frac{1}{3}(ab)(a^2+b^2).$$

(ii) $$\int_1^a\int_1^b\frac{dx\,dy}{xy} = \int_1^a\frac{1}{x}[\log y]_{y=1}^{b} dx,$$

(integrating w.r.t. y treating x as constant)

$$= \int_1^a\frac{(\log b-\log 1)}{x}dx$$

$$= \log b\int_1^a\frac{1}{x}dx = (\log b)[\log x]_1^a = (\log b)\,(\log a-\log 1)$$

$$= (\log b).(\log a).$$

(iii) $$\int_1^2\int_0^x\frac{dx\,dy}{x^2+y^2} = \int_1^2\left[\int_0^x\frac{dy}{x^2+y^2}\right]dx$$

$$= \int_1^2\left[\frac{1}{x}\tan^{-1}\frac{y}{x}\right]_{y=0}^{x} dx$$ (integrating w.r.t. y treating x as constant)

$$= \int_1^2\left[\frac{1}{x}(\tan^{-1}1-\tan^{-1}0)\right]dx = \frac{\pi}{4}\int_1^2\frac{dx}{x} = \frac{\pi}{4}[\log x]_1^2$$

$$= \frac{1}{4}\pi[\log 2-\log 1] = \frac{1}{4}\pi\log 2.$$

(iv) $$\int_0^{\pi/2}\int_{\pi/2}^{\pi}\cos(x+y)dy\,dx = \int_0^{\pi/2}\left[\int_{\pi/2}^{\pi}\cos(x+y)dx\right]dy$$

$$= \int_0^{\pi/2}[\sin(x+y)]_{x=\pi/2}^{\pi} dy,$$

(integrating w.r.t x treating y as constant)

$$= \int_0^{\pi/2}\left[\sin(\pi+y)-\sin\left(\frac{1}{2}\pi+y\right)\right]dy$$

$$= \int_0^{\pi/2} (-\sin y - \cos y) dy$$

$$= [\cos y - \sin y]_0^{\pi/2} = (0 - 1) - (1 - 0) = -2.$$

(v) $\int_0^1 \int_0^{x^2} e^{y/x} dx\, dy = \int_0^1 \left[x e^{y/x} \right]_{y=0}^{x^2} dx,$

(integrating w.r.t. y treating x as constant)

$$= \int_0^1 [x e^{x^2/x} - x e^{0/x}] dx = \int_0^1 (x e^x - x) dx$$

$$= \left[x e^x \right]_0^1 - \int_0^1 e^x dx - \left[\frac{x^2}{2} \right]_0^1$$

$$= e - \left[e^x \right]_0^1 - \frac{1}{2} = e - (e - 1) - \frac{1}{2} = 1 - \frac{1}{2} = \frac{1}{2}.$$

(vi) We have

$$\int_1^2 \int_0^{3y} dy\, dx = \int_1^2 y[x]_0^{3y} dy,$$

integrating w.r.t. x regarding y as a constant

$$= \int_1^2 y[3y - 0] dy = 3\int_1^2 y^2 dy = 3\left[\frac{y^3}{3} \right]_1^2 = \left[y^3 \right]_1^2 = 8 - 1 = 7.$$

(vii) The given integral

$$I = \int_{x=0}^{2} \int_{y=0}^{2x-4} \frac{2y - 1}{x + 1} dx\, dy$$

$$= \int_0^2 \frac{1}{x + 1} \left[y^2 - y \right]_{y=0}^{2x-4} dx,$$

integrating w.r.t. y treating x as constant

$$= \int_0^2 \frac{1}{x + 1} [(2x - 4)^2 - (2x - 4)] dx = \int_0^2 \frac{4x^2 - 18x + 20}{x + 1} dx$$

$$= \int_0^2 \left[4x - 22 + \frac{42}{x + 1} \right] dx, \qquad \text{dividing the Nr. by the Dr.}$$

$$= [2x^2 - 22x + 42 \log(x + 1)]_0^2$$

$$= 8 - 44 + 42 \log 3 = -36 + 42 \log 3.$$

Example 86:

Evaluate

(i) $\int_0^3\int_1^2 xy(1+x+y)dx\,dy.$

(ii) $\int_0^1\int_0^{\sqrt{(1+x^2)}} \frac{dx\,dy}{1+x^2+y^2}.$

(iii) $\int_0^2\int_0^{\sqrt{(4+x^2)}} \frac{dx\,dy}{4+x^2+y^2}.$

(iv) $\int_0^1\int_0^{\sqrt{(1-y^2)}} 4y\,dy\,dx.$

(v) $\int_0^1\int_x^{\sqrt{x}} (x^2-y^2)dx\,dy.$

(vi) $\int_2^3\int_0^{y-1} \frac{dy\,dx}{y}.$

Solution:

(i) $\int_0^3\int_1^2 xy(1+x+y)\,dx\,dy$

$$= \int_0^3\left[x\cdot\frac{y^2}{2}+x^2\cdot\frac{y^2}{2}+x\cdot\frac{y^3}{3}\right]_{y=1}^2 dx,$$

(integrating w.r.t. y treating x as constant)

$$= \int_0^3\left[\frac{x}{2}(4-1)+\frac{x^2}{2}(4-1)+\frac{x}{3}(8-1)\right]dx$$

$$= \int_0^3\left[\left(\frac{3}{2}+\frac{7}{3}\right)x+\frac{3}{2}x^2\right]dx = \left[\frac{23}{6}\cdot\frac{x^2}{2}+\frac{3}{2}\cdot\frac{x^3}{3}\right]_0^3$$

$$= \frac{23}{6}\cdot\frac{9}{2}+\frac{27}{2} = \frac{123}{4} = 30\frac{3}{4}.$$

(ii) $\int_0^1\int_0^{\sqrt{(1+x^2)}} \frac{dxdy}{1+x^2+y^2}$

$$= \int_0^1 \frac{1}{\sqrt{(1+x^2)}}\left[\tan^{-1}\frac{y}{\sqrt{(1+x^2)}}\right]_{y=0}^{\sqrt{(1+x^2)}} dx$$

(integrating w.r.t.y treating x as constant)

$$= \int_0^1 \frac{1}{\sqrt{(1+x^2)}}[\tan^{-1} 1 - \tan^{-1} 0]dx = \frac{\pi}{4}\int_0^1 \frac{dx}{\sqrt{(1+x^2)}}$$

$$= \frac{\pi}{4}\left[\log\{x + \sqrt{(1+x^2)}\}\right]_0^1 = \frac{\pi}{4}\log(1+\sqrt{2}).$$

(iii) The given integral

$$I = \int_{x=0}^{2}\int_{y=0}^{\sqrt{(4+x^2)}} \frac{dx\,dy}{(4+x^2)+y^2}$$

$$= \int_0^2 \frac{1}{\sqrt{(4+x^2)}}\left[\tan^{-1}\frac{y}{\sqrt{(4+x^2)}}\right]_{y=0}^{\sqrt{(4+x^2)}} dx,$$

(integrating w.r.t y treating x as constant)

$$= \int_0^2 \frac{1}{\sqrt{(4+x^2)}}[\tan^{-1} 1 - \tan^{-1} 0]dx = \frac{\pi}{4}\int_0^2 \frac{dx}{\sqrt{(4+x^2)}}$$

$$= \frac{\pi}{4}\left[\log\{x + \sqrt{(4+x^2)}\}\right]_0^2 = \frac{\pi}{4}[\log(2+2\sqrt{2}) - \log 2]$$

$$= \frac{\pi}{4}\log\frac{2+2\sqrt{3}}{2} = \frac{\pi}{4}\log(1+\sqrt{2}).$$

(iv) The given integral

$$I = \int_{y=0}^{1}\int_{x=0}^{\sqrt{(1-y^2)}} 4y\,dy\,dx$$

$$= \int_0^1 4y[x]_{x=0}^{\sqrt{(1-y^2)}} dy,$$ integrating w.r.t. x treating y as constant

$$= 4\int_0^1 y\sqrt{(1-y^2)}dy = 4\int_0^1\left(-\frac{1}{2}\right)\cdot(1-y^2)^{1/2}(-2y)dy$$

$$= -2\cdot\frac{2}{3}\left[(1-y^2)^{3/2}\right]_0^1,$$ by power formula

$$= -\frac{4}{3}[0-1] = \frac{4}{3}.$$

(v) The given integral

$$I = \int_{x=0}^{1}\int_{y=x}^{\sqrt{x}} (x^2+y^2)dx\,dy.$$

$$= \int_0^1 \left[x^2 y + \frac{1}{3}y^3\right]_{y=x}^{\sqrt{x}} dx,$$

integrating w.r.t. y treating x as constant

$$= \int_0^1 \left[x^2 \sqrt{x} + \frac{1}{3} x \sqrt{x} - x^3 - \frac{1}{3} x^3 \right] dx$$

$$= \int_0^1 \left[x^{5/2} + \frac{1}{3} x^{3/2} - \frac{4}{3} x^3 \right] dx$$

$$= \left[\frac{2}{7} x^{7/2} + \frac{1}{3} \cdot \frac{2}{5} x^{5/2} - \frac{1}{3} x^4 \right]_0^1$$

$$= \frac{2}{7} + \frac{2}{15} - \frac{1}{3} = \frac{30 + 14 - 35}{105} = \frac{3}{35}.$$

(vi) The given integral

$$I = \int_{y=2}^{3} \int_{x=0}^{y-1} \frac{dy\,dx}{y}$$

$$= \int_2^3 \frac{1}{y} [x]_{x=0}^{y-1} dy, \text{ integrating w.r.t. x treating y as constant}$$

$$= \int_2^3 \frac{y-1}{y} dy = \int_2^3 \left(1 - \frac{1}{y} \right) dy = [y - \log y]_2^3$$

$$= 3 - \log 3 - 2 + \log 2 = 1 - \log \frac{3}{2}.$$

Example 87:

Evaluate

(i) $\int_0^a \int_0^{\sqrt{(a^2 - y^2)}} \sqrt{(a^2 - x^2 - y^2)}\,dy\,dx.$

(ii) $\int_0^a \int_0^{\sqrt{(a^2 - y^2)}} (a^2 - x^2 - y^2)\,dy\,dx.$

(iii) $\int_0^a \int_0^{\sqrt{(a^2 - y^2)}} (x + y)\,dx\,dy.$

Solution:

(i) Here the variable limits are those of x and so the first integration must be performed w.r.t. x taking y as constant.

$$\therefore \int_0^a \int_0^{\sqrt{(a^2 - y^2)}} \sqrt{(a^2 - x^2 - y^2)}\,dy\,dx$$

$$= \int_0^a \left[\int_0^{\sqrt{(a^2 - y^2)}} \sqrt{\{(a^2 - y^2) - x^2\}}\,dx \right] dy$$

$$= \int_0^a \left[\frac{x\sqrt{(a^2 - y^2 - x^2)}}{2} + \frac{(a^2 - y^2)}{2} \sin^{-1} \frac{x}{\sqrt{(a^2 - y^2)}} \right]_{x=0}^{\sqrt{(a^2-y^2)}} dy$$

(integrating w.r.t. x treating y as constant)

$$= \int_0^a \left[0 + \frac{a^2 - y^2}{2} \cdot \frac{\pi}{2} \right] dy = \frac{\pi}{4} \left[a^2 y - \frac{y^3}{3} \right]_0^a = \frac{\pi}{4} \left[a^3 - \frac{a^3}{3} \right] = \frac{1}{6} \pi a^3.$$

(ii) The given integral

$$I = \int_{y=0}^{a} \int_{x=0}^{\sqrt{(a^2-y^2)}} [(a^2 - y^2) - x^2] dy\, dx$$

$$= \int_0^a \left[(a^2 - y^2)x - \frac{1}{3}x^3 \right]_{x=0}^{\sqrt{(a^2-y^2)}} dy,$$

Integrating w.r.t. x treating y as constant

$$= \int_0^a \left[(a^2 - y^2)^{3/2} - \frac{1}{3}(a^2 - y^2)^{3/2} \right] dy$$

$$= \frac{2}{3} \int_0^a (a^2 - y^2)^{3/2} dy$$

$$= \frac{2}{3} \int_0^{\pi/2} a^3 \cos^3 \theta .\ a \cos \theta\, d\theta,$$

putting $y = a \sin \theta$

so that $dy = a \cos \theta\, d\theta$

$$= \frac{2}{3} a^4 \int_0^{\pi/2} \cos^4 \theta\, d\theta$$

$$= \frac{2}{3} a^4 \cdot \frac{3.1}{4.2} \cdot \frac{\pi}{2}, \qquad \text{(by Walli's formula)}$$

$= \pi a4/8.$

(iii) The given integral

$$I = \int_{x=0}^{a} \int_{y=0}^{\sqrt{(a^2-x^2)}} (x + y) dx\, dy$$

$$= \int_0^a \left[xy + \frac{1}{2}y^2 \right]_{y=0}^{\sqrt{(a^2-x^2)}} dx,$$

integrating w.r.t. y treating x as constant

$$= \int_0^a \left[x\sqrt{(a^2 - x^2)} + \frac{1}{2}(a^2 - x^2) \right] dx$$

$$= \int_0^a \left[-\frac{1}{2}(a^2 - x^2)^{1/2}(-2x) + \frac{1}{2}(a^2 - x^2) \right] dx$$

$$= \left[-\frac{1}{2}\cdot\frac{2}{3}(a^2 - x^2)^{3/2} \right]_0^a + \frac{1}{2}\left[a^2x - \frac{1}{3}x^3 \right]_0^a, \quad \text{(by power formula)}$$

$$= 0 + \frac{1}{3}a^3 + \frac{1}{2}\left[a^3 - \frac{1}{3}a^3 \right] = \frac{2}{3}\ a^3.$$

Example 88:

Evaluate $\int_0^2 \int_0^{\sqrt{(2x-x^2)}} x\,dx\,dy$.

Solution:

Here the variable limits are those of y and so the first integration must be performed w.r.t. y regading x as constant.

$$\therefore \int_0^2 \int_0^{\sqrt{(2x-x^2)}} x\,dx\,dy = \int_0^2 x[y]_0^{\sqrt{(2x-x^2)}} dx$$

$$= \int_0^2 x\sqrt{(2x - x^2)}dx = \int_0^2 x\sqrt{\{1 - (1 - x)^2\}}dx.$$

Now put $(1 - x) = t$ so that $- dx = dt$.

Also when $x = 0$, $t = 1$ and when $x = 2$, $t = - 1$.

$$\therefore \quad \text{the required integral} = \int_{-1}^1 (1 - t)\sqrt{(1 - t^2)}dt$$

$$= \int_{-1}^1 \sqrt{(1 - t^2)}dt - \int_{-1}^1 t\sqrt{(1 - t^2)}dt$$

$$= 2\int_0^1 \sqrt{(1 - t^2)}dt - 0,$$

the second integral vanishes because the integrand is an odd function of t

$$= 2\left[\frac{t}{2}\sqrt{(1 - t^2)} + \frac{1}{2}\sin^{-1} t \right]_0^1$$

$$= 2\left[0 + \frac{1}{2}\cdot\frac{1}{2}\cdot\pi \right] = \frac{1}{2}\pi.$$

Example 89:

Evaluate $\int_0^1 \int_0^1 \frac{dx\,dy}{\sqrt{\{(1 - x^2)(1 - y^2)\}}}$.

Solution:

We have $\int_0^1\int_0^1 \frac{dx\,dy}{\sqrt{\{(1-x^2)(1-y^2)\}}}$

$$= \int_{y=0}^1 \frac{1}{\sqrt{(1-y^2)}}\left[\int_{x=0}^1 \frac{1}{\sqrt{(1-x^2)}}dx\right]dy$$

$$= \int_0^1 \frac{1}{\sqrt{(1-y^2)}}\left[\sin^{-1} x\right]_0^1 dy,$$

(integrating w.r.t. x treating y as constant)

$$= \int_0^1 \frac{\pi}{2\sqrt{(1-y^2)}}dy = \frac{\pi}{2}\left[\sin^{-1} y\right]_0^1 = \frac{\pi}{2}\cdot\frac{\pi}{2} = \frac{\pi^2}{4}.$$

Example 90:

Show that $\int_1^2\int_3^4 (xy+e^y)dx\,dy = \int_3^4\int_1^2 (xy+e^y)dx\,dy.$

Solution:

Integral on the L.H.S. $\int_1^2\left[\int_3^4 (xy+e^y)dy\right]dx$

$$= \int_1^2\left[\frac{xy^2}{2}+e^y\right]_3^4 dx = \int_1^2\left[8x+e^4-\frac{9}{2}x-e^3\right]dx$$

$$= \int_1^2\left[\frac{7}{2}x+e^4-e^3\right]dx = \left[\frac{7x^2}{4}+(e^4-e^3)x\right]_1^2$$

$$= 7 + 2(e^4-e^3) - \frac{7}{4} - (e^4-e^3) = \frac{21}{4} + e^4 - e^3.$$

And the integral on the R.H.S.

$$= \int_3^4\left[\int_1^2 (xy+e^y)dx\right]dy$$

$$= \int_3^4\left[\frac{yx^2}{2}+xe^y\right]_1^2 dy = \int_3^4\left[2y+2e^y-\frac{y}{2}-e^y\right]dy$$

$$= \int_3^4\left[\frac{3y}{2}+e^y\right]dy = \left[\frac{3y^2}{4}+e^y\right]_3^4$$

$$= 12 + e^4 - \frac{27}{4} - e^3 = \frac{21}{4} + e^4 - e^3.$$

Hence the result.

Example 91:

Show that $\int_1^2 \int_0^{y/2} y\,dy\,dx = \int_1^2 \int_0^{x/2} y\,dy\,dx.$

Solution:

We have

$$\int_1^2 \int_0^{y/2} y\,dy\,dx = \int_1^2 \left[y \int_0^{y/2} dx \right] dy = \int_1^2 y[x]_0^{Y/2} dy,$$

(integrating w.r.t. x treating y as a constant)

$$= \int_1^2 y\left[\frac{y}{2} - 0\right] dy = \frac{1}{2}\int_1^2 y^2 dy = \frac{1}{2}\left[\frac{y^3}{3}\right]_1^2 = \frac{1}{6}[8-1] = \frac{7}{6}$$

...(1)

Again $\int_1^2 \int_1^{x/2} x\,dx\,dy = \int_1^2 x\left[\int_0^{x/2} dy\right] dx = \int_1^2 x[y]_0^{x/2} dx,$

(integrating w.r.t. y treating x as a constant)

$$= \int_1^2 x\left[\frac{x}{2} - 0\right] dx = \frac{1}{2}\int_1^2 x^2 dx = \frac{1}{2}\left[\frac{x^3}{3}\right]_1^2 = \frac{1}{6}(8-1) = \frac{7}{6}$$

...(2)

Form (1) and (2), we see that

$$\int_1^2 \int_0^{y/2} y\,dy\,dx = \int_1^2 \int_0^{x/2} x\,dx\,dy.$$

Example 92:

Evaluate $\iint x^2y^2\,dx\,dy$ *over the region* $x^2 + y^2 \le 1$.

Solution:

Let R denote the region $x^2 + y^2 \le 1$. Then R is the region in the xy-plane bounded by the circle $x^2 + y^2 = 1$. The limits of integration for this region can be expressed either as $-1 \le x \le 1$, $-\sqrt{(1 - x^2)} \le y \le \sqrt{(1 - x^2)}$

or as $-\sqrt{(1 - y^2)} \le x \le \sqrt{(1 - y^2)}$, $-1 \le y \le 1$.

Because from the equation of the circle $x^2 + y^2 = 1$. we have $x^2=1-y^2$ so that $x = \pm\sqrt{(1 - y^2)}$. Thus for a fixed value of y, x varies from $-\sqrt{(1 - y^2)}$ to $\sqrt{(1 - y^2)}$ in the area bounded by the circle $x^2 + y^2 = 1$. Also y varies from -1 to 1 to cover the whole area of the circle $x^2 + y^2 = 1$. Therefore, if the first integration is to be performed w.r.t. x regarding y as constant, then

$$\iint_R x^2y^2 dx\,dy = \int_{y=-1}^{1}\int_{x=-\sqrt{(1-y^2)}}^{\sqrt{(1-y^2)}} x^2y^2 dx\,dy$$

$$= \int_{y=-1}^{1} y^2\left[\int_{x=-\sqrt{(1-y^2)}}^{\sqrt{(1-y^2)}} x^2 dx\right]dy$$

$$= \int_{-1}^{1} y^2\left[2\int_{x=0}^{\sqrt{(1-y^2)}} x^{2}dx\right]dy = \int_{-1}^{1} y^2\left[\frac{x^3}{3}\right]_0^{\sqrt{(1-y^2)}} dy$$

$$= \int_{-1}^{1}\frac{2}{3}y^2(1-y^2)^{3/2}dy = 2.\frac{2}{3}\int_0^1 2y^2(1-y^2)^{3/2}dy.$$

Put $y = \sin\theta$

so that $dy = \cos\theta\, d\theta$;

when $y = 0$, $\theta = 0$

and when $y = 1$, $\theta = \pi/2$.

$$\therefore \iint_R x^2y^2 dx\,dy = \frac{4}{3}\int_0^{\pi/2}\sin^2\theta(1-\sin^2\theta)^{3/2}.\cos\theta\, d\theta$$

$$= \frac{4}{3}\int_0^{\pi/2}\sin^2\theta\cos^4\theta\, d\theta = \frac{4}{3}\cdot\frac{1.3.1}{6.4.2}\cdot\frac{\pi}{2} = \frac{\pi}{24}.$$

Example 93:

Evaluate $\iint x^2y^3\, dx\, dy$ over the circle $x^2 + y^2 = a^2$.

Solution:

If the first integration is to be performed w.r.t. y regarding x as constant, then the region of integration R can be expressed as $- \le x \le a$, $-\sqrt{(a^2 - x^2)} \le y \le \sqrt{(a^2 - x^2)}$.

$$\therefore \quad \iint_R x^2y^3 dx\,dy = \int_{x=-a}^{a}\int_{y=-\sqrt{(a^2-x^2)}}^{\sqrt{(a^2-x^2)}} x^2y^3 dx\,dy = 0.$$

[$\because$ y^3 is an odd function of y]

Example 94:

Show that

$$\int_0^1 dx\int_0^1\frac{x-y}{(x+y)^3}dy \ne \int_0^1 dy\int_0^1\frac{x-y}{(x+y)^3}dx.$$

Find the values of the two integrals.

Solution:

The integral on the L.H.S.

$$= \int_0^1 dx \int_0^1 \frac{2x-(x+y)}{(+y)^3} dy$$

$$= \int_0^1 dx \int_0^1 \left\{ \frac{2x}{(x+y)^3} - \frac{1}{(x+y)^2} \right\} dy$$

$$= \int_0^1 \left[\frac{-x}{(x+y)^2} + \frac{1}{x+y} \right]_0^1 dx,$$

(integrating w.r.t. y regarding x as constant)

$$= \int_0^1 \left[-\frac{x}{(1+x)^2} + \frac{1}{x} + \frac{1}{1+x} - \frac{1}{x} \right] dx$$

$$= \int_0^1 \frac{dx}{(1+x)^2} = \left[\frac{-1}{1+x} \right]_0^1$$

$$= -\frac{1}{2} + 1 = \frac{1}{2}.$$

And the integral on the R.H.S.

$$= \int_0^1 dy \int_0^1 \frac{(x+y)-2y}{(x+y)^3} dx$$

$$= \int_0^1 dy \int_0^1 \left\{ \frac{1}{(x+y)^2} - \frac{2y}{(x+y)^3} \right\} dx$$

$$= \int_0^1 \left[\frac{-1}{x+y} + \frac{y}{(x+y)^2} \right]_0^1 dy$$

$$= \int_0^1 \left[\frac{-1}{1+y} + \frac{1}{y} + \frac{y}{(1+y)^2} - \frac{1}{y} \right] dy$$

$$= -\int_0^1 \frac{dy}{(1+y)^2} = \left[\frac{1}{1+y} \right]_0^1$$

$= \frac{1}{2} - 1 = -\frac{1}{2}$. Thus, the two integrals are not equal.

Example 95:

Find by double integration the area of the region bounded by the circle $x^2 + y^2 = a^2$.

Solution:

The area of a small element situated at any point (x, y) is dx dy. To find the area bounded by the circle $x^2 + y^2 = a^2$, the region of integration R can be expressed as $-a \le y \le a$, $-\sqrt{(a^2 - y^2)} \le x \le \sqrt{(a^2 - y^2)}$, where the first integration to be performed w.r.t. x regarding as constant.

∴ the required area

$$= \iint_R dx\,dy = \int_{y=-a}^{a}\int_{x=-\sqrt{(a^2-y^2)}}^{\sqrt{(a^2-y^2)}} 1.dx\,dy$$

$$= \int_{-a}^{a}\left[2\int_{0}^{\sqrt{(a^2-y^2)}} 1.dx\right]dy = 2\int_{-a}^{a}[x]_0^{\sqrt{(a^2-y^2)}}\,dy$$

$$= 2\int_{-a}^{a}\sqrt{(a^2 - y^2)}dy = 2.2\int_0^a\sqrt{(a^2 - y^2)}dy$$

$$= 4\left[\frac{y\sqrt{(a^2 - y^2)}}{2} + \frac{a^2}{2}\sin^{-1}\frac{y}{a}\right]_0^a$$

$$= 4\left[0 + \frac{a^2}{2}\sin^{-1} 1\right]$$

$$= 4.\frac{1}{2}a^2.\frac{1}{2}\pi = \pi a^2$$

2

Beta and Gamma Functions

(Euler's Integrals, Dirichlet's Integrals, Liouville's Extension of Dirichliet's Theorem)

2.1 Beta Functions

Definition

The definite integral

$$\int_0^1 x^{m-1}(1-x)^{n-1}\,dx, \textit{ for } m > 0,\ n > 0$$

is called the ***Beta function*** *and is denoted by* **B** *(m, n) [read as "Beta m, n"]. Thus*

$$B\,(m, n) = \int_0^1 x^{m-1}(1-x)^{n-1}\,dx,$$

where m, n are positive numbers, integral or fractional. Beta function is also called the *Eulerian integral of the first kind.*

2.2 Some Simple Properties of Beta Functions

(i) Symmetry of Beta function i.e. B (m, n) = B (n.m).

We have $B\,(m, n) = \int_0^1 x^{m-1}(1-x)^{n-1}\,dx$, dy the def. of the Beta function

$$= \int_0^1 (1-x)^{m-1}\{1-(1-x)\}^{n-1}\,dx, \qquad \left[\because \int_0^a (x)dx = \int_0^a f(a-x)dx\right]$$

$$= \int_0^1 (1-x)^{m-1} x^{n-1}dx = \int_0^1 x^{n-1}(1-x)^{m-1}\,dx$$

= B (n, m), by the def. of Beta function.

Hence **B** *(m, n)* = **B** *(n, m).*

(ii) If m or n is a positive integral, B (m, n) can be evaluated in an application form.

Case I:

When n is a positive integer. If n = 1, the result is obvious because **B** (m, 1) $= \int_0^1 x^{m-1}(1-x)^{1-1}\,dx = \int_0^1 x^{m-1}dx$

$$= \left[\frac{x^m}{m}\right]_0^1 = \frac{1}{m}.$$

So let us take n > 1. We have

$$\mathbf{B}\,(m, n) = \int_0^1 x^{m-1}(1-x)^{n-1}\,dx = \int_0^1 (1-x)^{n-1}x^{m-1}dx$$

$$= \left[(1-x)^{n-1}\cdot\frac{x^m}{m}\right]_0^1 - \int_0^1 (n-1)(1-x)^{n-2}(-1)\cdot\frac{x^m}{m}dx$$

integrating by parts taking x^{m-1} as the second function

$$= 0 + \frac{n-1}{m}\cdot\int_0^1 x^m(1-x)^{n-2}\,dx, \qquad [\because n > 1]$$

$$= \frac{n-1}{m}\cdot\int_0^1 x^{(m+1)-1}(1-x)^{(n-1)-1}\,dx$$

$$= \frac{n-1}{m}B(m+1, n-1).$$

By the repeated application of this process, we get

$$B\,(m, n) = \frac{n-1}{m}\cdot\frac{n-2}{m+1}\cdot\frac{n-3}{m+2}\cdots\frac{1}{m+n-2}B(m+n-1, 1)$$

$$= \frac{n-1}{m}\cdot\frac{n-2}{m+1}\cdot\frac{n-3}{m+2}\cdots\frac{1}{m+n-2}\int_0^1 x^{m+n-2}(1-x)^0\,dx$$

$$= \frac{n-1}{m}\cdot\frac{n-2}{m+1}\cdot\frac{n-3}{m+2}\cdots\frac{1}{m+n-2}\int_0^1 x^{m+n-2}dx$$

$$= \frac{(n-1)!}{m(m+1)(m+2)\ldots(m+n-2)}\cdot\left[\frac{x^{m+n-1}}{m+n-1}\right]_0^1.$$

$$\therefore \mathbf{B}\,(m, n) = \frac{(n-1)!}{m(m+1)(m+2)\ldots(m+n-2)(m+n-1)}.$$

Case II:

When m is a positive integer. Since the Beta function is symmetrical in m and n *i.e.*, **B** (m, n) = **B** (n, m), therefore by case I, we conclude that

$$B(m, n) = \frac{(n-1)!}{n(n+1)(n+2)...(n+m-2)(n+m-1)}.$$

(iii) If both m and n are positive integers, then

$$\mathbf{B(m, n)} = \frac{(m-1)!(n-1)!}{(m+n-1)!}.$$

From (ii), we have

$$B(m, n) = \frac{(n-1)!}{m(m+1)(m+2)\cdots(m+n-2)(m+n-1)}.$$

$$= \frac{(n-1)!(m-1)!}{(m+n-1)(m+n-2)\cdots(m+1)m(m-1)!},$$

writing the denominator in the reversed order and multiplying the Nr and Dr by (m – 1) !

$$= \frac{(m-1)!(n-1)!}{(m+n-1)!}.$$

2.3 Another Form of Beta Function

$$\mathbf{B(m, n)} = \int_0^\infty \frac{x^{m-1}}{(1+x)^{m+n}} dx, \ \mathbf{m > 0, n > 0.}$$

Proof:

By the definition of Beta function, we have

$$B(m, n) = \int_0^1 x^{m-1}(1-x)^{n-1} dx.$$

Put $x = \frac{1}{1+y}$

so that $dx = -\frac{1}{(1+y)^2} dx$.

Also when $x \to 0$, $y \to \infty$

and when $x = 1$, $y = 0$.

$$\therefore B(m, n) = \int_\infty^0 \frac{1}{(1+y)^{m-1}} \cdot \left[1 - \frac{1}{1+y}\right]^{n-1} \cdot \left[-\frac{1}{(1+y)^2}\right] dy$$

$$= \int_0^\infty \frac{1}{(1+y)^{m-1+2}} \cdot \frac{y^{n-1}}{(1+y)^{n-1}} dy = \int_0^\infty \frac{y^{n-1}}{(1+y)^{m+n}} dy$$

$$= \int_0^\infty \frac{x^{n-1}}{(1+y)^{m+n}} dy,$$ (by a property of definite integrals)

...(1)

Again since Beta function is symmetrical in m and , we have

$$\mathbf{B}(m, n) = \mathbf{B}(n, m) = \int_0^\infty \frac{x^{m-1}}{(1+y)^{m+n}} dy, \qquad \text{by (1).}$$

$$\text{Thus } \mathbf{B}(m, n) = \int_0^\infty \frac{x^{m-1}}{(1+y)^{m+n}} dy = \int_0^\infty \frac{x^{n-1}}{(1+y)^{m+n}} dy, \quad m > 0, n > 0.$$

2.4 Gamma Functions

Definition : The definite integral

$$\int_0^\infty e^{-x} x^{n-1} dx, \textit{ for } n > 0$$

is called the Gamma Function and is denoted by $\Gamma(n)$ (read as "Gamma n"). Thus

$$\Gamma(n) = \int_0^\infty e^{-x} x^{n-1} dx, \text{ for } n > 0.$$

Gamma function is also called *Eulerian integral of the second find.*

2.5 Fundamental Property of Gamma Functions

To prove that

(i) $\Gamma(n+1) = n\,\Gamma(n)$, *where* $n > 0$

and (ii) $\Gamma(n) = (n-1)!$, *where n is a positive integer.*

Proof:

By the definition of gamma function, we have

$$\Gamma(n+1) = \int_0^\infty e^{-x} x^{(n+1)-1} dx = \int_0^\infty x^n e^{-x} dx$$

$$= \left[-e^{-x} x^n\right]_0^\infty + \int_0^\infty e^{-x} . nx^{n-1} dx, \qquad \text{...(1)}$$

integrating by parts taking e^{-x} as the second function.

$$\text{Now } \lim_{x \to \infty} \frac{x^n}{e^x} = \lim_{x \to \infty} \frac{x^n}{1 + x + (x^2/2!) + \ldots + (x^n/n!) + \ldots}$$

$$= \lim_{x \to \infty} \frac{1}{\frac{1}{x^n} + \frac{1}{x^{n-1}} + \ldots + \frac{1}{n!} + \frac{x}{(n+1)!} + \ldots} = \frac{1}{\infty} = 0.$$

$\therefore$ from (1), we get

$$\Gamma(n+1) = 0 + n \int_0^\infty e^{-x} x^{n-1} dx, \qquad [\because n > 0]$$

$= n\,\Gamma(n)$, which proves the result (i).

(ii) We have $\Gamma(n) = \Gamma[(n-1)+1] = (n-1)\,\Gamma(n-1)$,

$[\because \Gamma(n+1) = n\,\Gamma(n)]$.

Similarly $\Gamma(n-1) = (n-2)\,\Gamma(n-2)$, ... etc.

Hence if n is a +ive integer, then proceeding as above, we get

$\Gamma(n) = (n-1)(n-2) \ldots 2.1\,\Gamma(1)$.

But $\Gamma(1) = \int_0^\infty e^{-x} x^{1-1} dx = \int_0^\infty e^{-x} x^0 dx = \int_0^\infty e^{-x}.1dx$

$$= \left[\frac{e^{-x}}{-1}\right]_0^\infty = -\left[\lim_{x\to\infty} \frac{1}{e^x} - e^0\right] = -[0-1] = 1.$$

Hence $\Gamma(n) = (n-1)(n-2) \ldots 2.1.1 = (n-1)!$

if n is a + ive integer.

Remember

$\Gamma(n) = (n-1)\,\Gamma(n-1)$, where $n > 1$ and $\Gamma(1) = 1$.

Also it may be remarked that $\Gamma(0) = \infty$ and $\Gamma(-n) = \infty$ where n is a positive integer.

2.6 Some Transformations of Gamma Functions

We have $\Gamma(n) = \int_0^\infty x^{n-1} e^{-x} dx.$

(i) put $x = ay$

so that $dx = a\,dy$;

when $x = 0$, $y = 0$

and when $x \to \infty$,

$y \to \infty$.

$$\therefore \Gamma(n) = \int_0^\infty e^{-ay} a^n y^{n-1} dy.$$

Hands $\int_0^\infty e^{-ay} y^{n-1} dy = \frac{\Gamma(n)}{a^n}.$

(ii) In (1) if we put $x = \log(1/y)$

or $y = e^{-x}$ so that

$dy = -e^{-x} dx$,

Then $\Gamma(n) = -\int_1^0 \left(\log\frac{1}{y}\right)^{n-1} dy. = \int_0^1 \left(\log\frac{1}{y}\right)^{n-1} dy.$

(iii) In (1) if we put $x^n = y$

so that $nx^{n-1}\,dx = dy$, we get

$$\Gamma(n) = \frac{1}{n}\int_0^\infty e^{-(y)^{1/n}}dy$$

or $$\frac{1}{n}\int_0^\infty e^{-(y)^{1/n}}dy = n\,\Gamma(n) = \Gamma(n+1).$$

2.7 Relation Between Beta and Gamma Functions

To prove that

$$\mathbf{B(m, n) = \frac{\Gamma(m)\Gamma(n)}{\Gamma(m+n)}, \text{ where } m > 0, n > 0.}$$

Proof:

We have

$$\frac{\Gamma(m)}{z^m} = \int_0^\infty e^{-zx}x^{m-1}dx.$$

$$\therefore \Gamma(m) = z^m\int_0^\infty e^{-zx}.x^{m-1}dx = \int_0^\infty z^m e^{-zx}x^{m-1}dx.$$

Multiplying both sides by $e^{-z}x^{n-1}$, we get

$$\Gamma(m)e^{-z}z^{n-1} = \int_0^\infty e^{-z}(1+x)z^{m+n-1}x^{m-1}dx. \qquad ...(1)$$

Now integrating both sides of (1) with respect to z from 0 to ∞, we get

$$\Gamma(m)\int_0^\infty e^{-z}z^{n-1}dz = \int_0^\infty\left[\int_0^\infty e^{-z(1+x)}z^{m+n-1}x^{m-1}dx\right]dz$$

$$\text{or } \Gamma(m)\Gamma(n) = \int_0^\infty\left[\int_0^\infty e^{-z(1+x)}z^{m+n-1}dz\right]x^{m-1}dx$$

$$= \int_0^\infty \frac{\Gamma(m+n)}{(1+x)^{m+n}}z^{m-1}dx, \qquad \text{[by 6, part (ii)]}$$

$$= \Gamma(m+n)\int_0^\infty \frac{x^{m-1}}{(1+x)^{m+n}}x^{m-1}dx$$

$= \Gamma(m+n).\,B(m, n)$, by 3.

$$\therefore B(m, n) = \frac{\Gamma(m)\Gamma(n)}{\Gamma(m+n)}.$$

Thus remember that

$$\int_0^1 x^{m-1}(1-x)^{n-1}dx = \frac{\Gamma(m)\Gamma(n)}{\Gamma(m+n)}.$$

Cor. : $\boldsymbol{\Gamma(n)\Gamma(1-n) = \dfrac{\pi}{\sin n\pi}}$**, where** $\boldsymbol{0 < n < 1}$**.**

Proof:

We know that $B(m, n) = \int_0^\infty \frac{x^{n-1}}{(1+x)^{m+n}}dx$, [See 3]

and $B(m, n) = \frac{\Gamma(m)\Gamma(n)}{\Gamma(m+n)}$, where $m > 0$ and $n > 0$.

$$\therefore \frac{\Gamma(m)\Gamma(n)}{\Gamma(m+n)} = \int_0^\infty \frac{x^{n-1}}{(1+x)^{m+n}}dx.$$

Putting $m + n = 1$

or $m = 1 - n$ in the above relation, we get

$$\frac{\Gamma(1-n)\Gamma(n)}{\Gamma(1)} = \int_0^\infty \frac{x^{n-1}}{1+x}dx, \text{ where } 0 < n < 1.$$

[Note that $m > 0 \Rightarrow 1 - n > 0 \Rightarrow n < 1$.]

But $\Gamma(1) = 1$. Also

$$\int_0^\infty \frac{x^{n-1}}{1+x}dx = \frac{\pi}{\sin n\pi}.$$

$$\therefore \Gamma(n)\Gamma(1-n) = \frac{\pi}{\sin n\pi}, \text{ where } 0 < n < 1.$$

2.8 The value of Γ (1/2)

To prove that $\Gamma\left(\frac{1}{2}\right) = \sqrt{\pi}$.

Proof:

We know that $B(m, n) = \frac{\Gamma(m)\Gamma(n)}{\Gamma(m+n)}$ ...(1)

If we take $m = \frac{1}{2}$, $n = \frac{1}{2}$, then from (1), we have

$$B\left(\frac{1}{2}, \frac{1}{2}\right) = \frac{\Gamma\left(\frac{1}{2}\right)\Gamma\left(\frac{1}{2}\right)}{\Gamma\left(\frac{1}{2}+\frac{1}{2}\right)} = \frac{\left[\Gamma\left(\frac{1}{2}\right)\right]^2}{\Gamma(1)} = \left[\Gamma\left(\frac{1}{2}\right)\right]^2 . [\because \Gamma(1) = 1]$$

Thus $\left[\Gamma\left(\frac{1}{2}\right)\right]^2 = B\left(\frac{1}{2}, \frac{1}{2}\right)$

$= \int_0^1 x^{1/2-1}(1-x)^{1/2-1}\,dx,$

by the definition of Beta function

$= \int_0^1 x^{-1/2}(1-x)^{-1/2}\,dx.$

Now put $x = \sin^2\theta$ so that $dx = 2\sin\theta\cos\theta\,d\theta$.

Also when $x = 0$, $\theta = 0$ and when $x = 1$, $\theta = \frac{1}{2}\pi$.

$\therefore \left[\Gamma\left(\frac{1}{2}\right)\right]^2 = \int_0^{\pi/2} \frac{1}{\sin\theta}\cdot\frac{1}{\cos\theta}\cdot 2\sin\theta\cos\theta\,d\theta$

$= 2\int_0^{\pi/2} d\theta = 2\,[\theta]_0^{\pi/2}$

$= 2\left[\frac{1}{2}\pi - 0\right] = \pi.$

Taking square root of both the sides, we get

$$\Gamma\left(\frac{1}{2}\right) = \sqrt{\pi}.$$

Important Deduction

To prove that $\int_0^\infty e^{-x^2}\,dx = \frac{\sqrt{\pi}}{2}$.

Proof:

Let $I = \int_0^\infty e^{-x^2}\,dx$.

Put $x^2 = z$

so that $2x\,dx = dz$

or $dx = \frac{1}{2x}dz = \frac{1}{2\sqrt{z}}dz = \frac{1}{2}z^{-1/2}dz$.

Also when $x = 0$, $z = 0$

and when $x \to \infty$, $z \to \infty$.

$\therefore I = \int_0^\infty e^{-z}\frac{1}{2}z^{-1/2}dz = \frac{1}{2}\int_0^\infty e^{-z}z^{1/2-1}dz = \frac{1}{2}\Gamma\left(\frac{1}{2}\right) = \frac{\sqrt{\pi}}{2}.$

Hence $\int_0^{\infty} e^{-x^2}\, dx = \frac{\sqrt{\pi}}{2}$.

2.9 To Prove That for All Values of m and n Such That m > –1, > –1

$$\int_0^{\pi/2} \cos^m\theta \sin^n\theta\, d\theta = \frac{\Gamma\left(\frac{m+1}{2}\right)\Gamma\left(\frac{n+1}{2}\right)}{2\Gamma\left(\frac{m+n+2}{2}\right)}$$

Proof:

put $\sin^2\theta = x$

so that $2 \sin\theta \cos\theta\, d\theta = dx$

or $2 \sin\theta. \sqrt{(1 - \sin^2\theta)}\, d\theta = dx$

or $2x^{1/2}.\sqrt{(1 - x)}\, d\theta = dx$.

$$\therefore d\theta = \frac{dx}{2x^{1/2}(1-x)^{1/2}}.$$

Also when $\theta = 0$, $x = 0$ and when $\theta = \frac{1}{2}\pi$, $x = 1$.

$$\therefore \int_0^{\pi/2} \text{xos}^m\theta \sin^n\theta\, d\theta = \int_0^{\pi/2} (1-\sin^2\theta)^{m/2}.\sin^n\theta\, d\theta$$

$$= \int_0^1 (1-x)^{m/2}.x^{n/2}.\frac{dx}{2x^{1/2}(1-x)^{1/2}}$$

$$= \frac{1}{2}\int_0^1 x^{(n-1)/2}(1-x)^{(m-1)/2}\, dx$$

$$= \frac{1}{2}\int_0^1 x^{\{(n-1)/2\}-1}(1-x)^{\{(m-1)/2\}-1}\, dx$$

$$= \frac{1}{2}B\left(\frac{m+1}{2}, \frac{n+1}{2}\right), \qquad \text{(provided } m > -1 \text{ and } n > -1)$$

$$= \frac{1}{2}\frac{\Gamma\frac{1}{2}(m+1)\Gamma\frac{1}{2}(n+1)}{\Gamma\frac{1}{2}(m+1+n+1)}, \qquad \left[\because B(m,n) = \frac{\Gamma(m)\Gamma(n)}{\Gamma(m+n)}\right]$$

$$= \frac{\Gamma\frac{1}{2}(m+1)\Gamma\frac{1}{2}(n+1)}{2\Gamma\frac{1}{2}(m+n+2)}.$$

2.10 Some Important Transformations of Beta Function

Beta function can be transformed into many other forms. A few of them are given below.

(i) We know that $\int_0^{\infty} \frac{y^{m-1}}{(1+y)^{m+n}} dy = B(m, n)$.

Now $\int_0^{\infty} \frac{y^{m-1}}{(1+y)^{m+n}} dy = \int_0^{1} \frac{y^{m-1}}{(1+y)^{m+n}} dy + \int_0^{\infty} \frac{y^{m-1}}{(1+y)^{m+n}} dy$.

Making the substitution y = 1/x in the last integral, we get

$$\int_0^{\infty} \frac{y^{m-1}}{(1+y)^{m+n}} dy = \int_0^{1} \frac{x^{n-1} dx}{(1+x)^{m+n}}.$$

$$\therefore B(m, n) = \int_0^{\infty} \frac{y^{m-1}}{(1+y)^{m+n}} dy$$

$$= \int_0^{1} \frac{y^{m-1}}{(1+y)^{m+n}} dy + \int_0^{1} \frac{x^{n-1}}{(1+y)^{m+n}} dx$$

$$= \int_0^{1} \frac{x^{m-1}}{(1+x)^{m+n}} dx + \int_0^{1} \frac{x^{n-1}}{(1+x)^{m+n}} dx$$

$$= \int_0^{1} \frac{x^{m-1} + x^{n-1}}{(1+y)^{m+n}} dx$$

Hence $\int_0^{1} \frac{x^{m-1} + x^{n-1}}{(1+y)^{m+n}} dx = \mathbf{B(m, n)} = \frac{\Gamma(\mathbf{m})\Gamma(\mathbf{n})}{\Gamma(\mathbf{m}+\mathbf{n})}$.

(ii) We know that $\int_0^{\infty} \frac{x^{m-1}}{(1+x)^{m+x}} dx = \mathbf{B}(m, n)$.

If we put $x = \frac{ay}{b}$,

so that $dx = \frac{a}{b} dy$, we get

$$\int_0^{\infty} \frac{x^{m-1}}{(1+x)^{m+x}} dx = a^m b^n \int_0^{\infty} \frac{y^{m-1}}{(ay+b)^{m+n}} dy.$$

$$\therefore \int_0^{\infty} \frac{y^{m-1}}{(ay+b)^{m+n}} dy = \frac{1}{a^m b^n} \int_0^{\infty} \frac{x^{m-1}}{(1+x)^{m+n}} dx$$

$$= \frac{1}{a^m b^n} \mathbf{B}(m, n).$$

Hence $\int_0^{\infty} \frac{y^{m-1}}{(ay+b)^{m+n}} dy = \frac{\Gamma(m)\Gamma(n)}{a^m b^n \Gamma(m+n)}$.

Again putting $y = \tan^2\theta$ *i.e.*, $dy = 2\tan\theta\sec^2\theta\, d\theta$ in the integral just obtained, we get

$$\int_0^{\pi/2} \frac{\sin^{2m-1}\theta\cos^{2n-1}\theta\, d\theta}{(a\sin^2\theta + b\cos^2\theta)^{m+n}} = \frac{\Gamma(m).\Gamma(n)}{2a^m b^n \Gamma(m+n)}.$$

(iii) We know that $\int_0^1 x^{m-1}(1-x)^{n-1} dx = \mathbf{B}(m, n)$.

Putting $x = \sin^2\theta$,

so that $dx = 2\sin\theta\cos\theta\, d\theta$, we have

$$\int_0^1 x^{m-1}(1-x)^{n-1} dx = 2\int_0^{\pi/2} \sin^{2m-1}\theta\cos^{2n-1}\theta\, d\theta.$$

$$\int_0^{\pi/2} \sin^{2m-1}\theta\cos^{2n-1}\theta\, d\theta = \frac{\mathbf{B}(m,n)}{2} = \frac{\Gamma(m)\Gamma(n)}{2\Gamma(m+n)}.$$

This result may also be written in the form

$$\int_0^{\pi/2} \sin^p\theta\cos^q\theta\, d\theta = \frac{\Gamma\left(\frac{p+1}{2}\right)\cdot\Gamma\left(\frac{q+1}{2}\right)}{2\Gamma\left(\frac{p+q+2}{2}\right)},$$

by putting $2m - 1 = p$ and $2n - 1 = q$.

(iv) We have that $\int_0^1 y^{m-1}(1-y)^{n-1} dy = \mathbf{B}(m, n)$.

Putting $y = \frac{x-b}{a-b}$,

so that $dy = \frac{dx}{a-b}$, we have

$$\int_0^1 y^{m-1}.(1-y)^{n-1} dy = \int_b^a \left(\frac{x-b}{a-b}\right)^{m-1}\left(\frac{a-x}{a-b}\right)^{n-1} \cdot \frac{dx}{a-b}$$

$$= \frac{1}{(a-b)^{m+n-1}} \int_b^a (x-b)^{m-1}(a-x)^{n-1} dx.$$

$$\therefore \int_b^a (x-b)^{m-1}(a-x)^{n-1} dx$$

$$= (a-b)^{m+n-1} \int_0^1 y^{m-1}(1-y)^{n-1} dy$$

$$\int_b^a (x-b)^{m-1}(a-x)^{n-1}\,dx = (a-b)^{m+n-1} B(m,n)$$

$$= (a-b)m+n-1\ \frac{\Gamma(m)\Gamma(n)}{\Gamma(m+n)}.$$

2.11 Duplication Formula

To prove that $\Gamma(m)\,\Gamma\left(m+\frac{1}{2}\right) = \frac{\sqrt{\pi}}{2^{2m-1}}\Gamma(2m)$, where $m > 0$.

Proof:

We know that

$$B(m, n) = \frac{\Gamma(m)\Gamma(n)}{\Gamma(m+n)}, \text{ where } m > 0, n > 0.$$

If we take n = m, then

$$B(m, m) = \frac{[\Gamma(m)]^2}{\Gamma(2m)}. \qquad \ldots(1)$$

Again by the definition of Beta function, we have

$$\mathbf{B}(m, m) = \int_0^1 x^{m-1}(1-x)^{m-1}\,dx.$$

Let us put $x = \sin^2\theta$

so that $dx = 2\sin\theta\cos\theta\,d\theta$.

Also when $x = 0$, $\theta = 0$

and when $x = 1$, $\theta = \frac{1}{2}\pi$.

Then $\mathbf{B}(m, m) = \int_0^{\pi/2} \sin^{2(m-1)}\theta.\cos^{2(m-1)}\theta.2\sin\theta\cos\theta\,d\theta$

$$= 2\int_0^{\pi/2}\sin^{2m-1}\theta.\cos^{2m-1}\theta\,d\theta = 2\int_0^{\pi/2}(\sin\theta\cos\theta)^{2m-1}\,d\theta$$

$$= 2\int_0^{\pi/2}\left(\frac{\sin 2\theta}{2}\right)^{2m-1}d\theta = \frac{1}{2^{2m-2}}\int_0^{\pi/2}\sin^{2m-1}2\theta\,d\theta$$

$$= \frac{1}{2^{2m-2}}\int_0^{\pi}\sin^{2m-1}\phi\cdot\frac{d\phi}{2}, \text{ putting } 2\theta = \phi \text{ so that } d\theta = \frac{1}{2}d\phi$$

$$= \frac{1}{2^{2m-1}}\int_0^{\pi}\sin^{2m-1}\phi\,d\phi = \frac{1}{2^{2m-1}}\cdot 2\int_0^{\pi/2}\sin^{2m-1}\phi\,d\phi$$

$$= \frac{1}{2^{2m-2}}\int_0^{\pi/2}\sin^{2m-1}\phi.\cos^0\phi\,d\phi$$

$$= \frac{1}{2^{2m-2}} \frac{\Gamma\frac{1}{2}(2m-1+1)\Gamma\frac{1}{2}(0+1)}{2\Gamma\frac{1}{2}(2m-1+0+2)} = \frac{1}{2^{2m-1}} \cdot \frac{\Gamma(m)\Gamma\left(\frac{1}{2}\right)}{\Gamma\left(m+\frac{1}{2}\right)}$$

$$= \frac{1}{2^{2m-1}} \cdot \frac{\Gamma(m)\sqrt{\pi}}{\Gamma\left(m+\frac{1}{2}\right)}. \qquad \left[\because \Gamma\left(\frac{1}{2}\right) = \sqrt{\pi}\right] \quad ...(2)$$

Now equating the two values of B (m, m) obtained in (1) and (2), we get

$$\frac{[\Gamma(m)^2}{\Gamma(2m)} = \frac{1}{2^{2m-1}} \cdot \frac{\Gamma(m).\sqrt{\pi}}{\Gamma\left(m+\frac{1}{2}\right)}$$

or $\Gamma(m)\,\Gamma\left(m+\frac{1}{2}\right) = \frac{\sqrt{\pi}}{2^{2m-1}}\Gamma(2m)$.

2.12 To Find the Value of

$$\boldsymbol{\Gamma\left(\frac{1}{n}\right)\Gamma\left(\frac{2}{n}\right)\Gamma\left(\frac{3}{n}\right)\ldots\Gamma\left(\frac{n-1}{n}\right)}$$

where n is a positive integer.

Proof:

Let $A = \Gamma\left(\frac{1}{n}\right)\Gamma\left(\frac{2}{n}\right)\Gamma\left(\frac{3}{n}\right)\ldots\Gamma\left(\frac{n-1}{n}\right)$. ...(1)

Writing the above expression in the reverse order, we have

$$A = \Gamma\left(1-\frac{1}{n}\right)\Gamma\left(1-\frac{2}{n}\right)\ldots\Gamma\left(1-\frac{n-1}{n}\right)\Gamma\left(1-\frac{n-1}{n}\right). \quad ...(2)$$

Multiplying (1) and (2), we get

$$A^2 = \Gamma\left(\frac{1}{n}\right)\Gamma\left(1-\frac{1}{n}\right)\Gamma\left(\frac{2}{n}\right)\Gamma\left(\frac{2}{n}\right)\ldots\Gamma\left(\frac{n-1}{n}\right)\Gamma\left(1-\frac{n-1}{n}\right)$$

$$= \frac{\pi^{n-1}}{\sin\frac{\pi}{n}\sin\frac{2\pi}{n}\ldots\sin\frac{n-1}{n}\pi}.$$

To calculate this expression, we factorize $1 - x^{2n}$.

Now the roots of the equation $x^{2n} - 1 = 0$ are given by

$$x = (1)^{1/2n} = (\cos 2r\pi + i\sin 2r\pi)^{1/2n}$$

$$= \cos\frac{r\pi}{n} + i\sin\frac{r\pi}{n}, \text{ where } r = 0,1,2,\ldots,2n-1.$$

Hence, we have $1 - x^{2n}$

$$= (1-x)(1+x)\left(x-\cos\frac{\pi}{n}-i\sin\frac{\pi}{n}\right)\left(x-\cos\frac{\pi}{n}+i\sin\frac{\pi}{n}\right)\ldots$$

$$\ldots\left(x-\cos\frac{n-1}{n}\pi-i\sin\frac{n-1}{n}\pi\right)\left(x-\cos\frac{n-1}{n}\pi+i\sin\frac{n-1}{n}\pi\right)$$

$$= (1-x^2)\left(1-2x\cos\frac{\pi}{n}+x^2\right)\left(1-2x\cos\frac{2\pi}{n}+x^2\right)$$

$$\ldots\left(1-2x\cos\frac{n-1}{n}\pi+x^2\right).$$

$$\frac{1-x^{2n}}{1-x^2} = \left(1-2x\cos\frac{\pi}{n}+x^2\right)\left(1-2x\cos\frac{2\pi}{n}+x^2\right)\ldots$$

$$\ldots\left(1-2x\cos\frac{n-1}{n}\pi+x^2\right)$$

putting $x = 1$ and $x = -1$ respectively, we have in the limit,

$$n = \left(2-2\cos\frac{\pi}{n}\right)\left(2-2\cos\frac{2\pi}{n}\right)\ldots\left(2-2\cos\frac{n-1}{n}\pi\right).$$

and $$n = \left(2+2\cos\frac{\pi}{n}\right)\left(2+2\cos\frac{2\pi}{n}\right)\ldots\left(2+2\cos\frac{n-1}{n}\pi\right).$$

Multiplying these, we get

$$n^2 = 2^{2n-2}\sin^2\frac{\pi}{n}\ldots\sin^2\frac{n-1}{n}\pi$$

or $n = 2^{n-1}.\sin\frac{\pi}{n}\ldots\sin\frac{n-1}{n}\pi.$

Hence, from (3), we get

$$A^2 = \frac{\pi^{n-1}}{n/2^{n-1}} = \frac{(2\pi)^{n-1}}{n}$$

or $A = \dfrac{(2\pi)^{(n-1)/2}}{n^{1/2}}.$

Remark:

The value of $\sin\frac{\pi}{n}\sin\frac{2\pi}{n}\ldots\sin\frac{n-1}{n}\pi$ can also be found by using the trigonometrical identity

$$\frac{\sin n\theta}{\sin\theta} = 2^{n-1}\sin\left(\theta+\frac{\pi}{n}\right)\sin\left(\theta+\frac{2\pi}{n}\right)\sin\left(\theta+\frac{3\theta}{n}\right)\ldots\sin\left(\theta+\frac{n-1}{n}\pi\right).$$

From the above identity, we have

$$\frac{\sin n\theta}{n\theta}\cdot\frac{\theta}{\sin\theta}.n = 2^{n-1}\sin\left(\theta+\frac{\pi}{n}\right)\sin\left(\theta+\frac{2\pi}{n}\right)\ldots\ldots\sin\left(\theta+\frac{n-1}{n}p\right)$$

Taking limit as $\theta \to 0$, we get

$$n = 2^{n-1}\sin\frac{\pi}{n}\sin\frac{2\pi}{n}\sin\frac{3\pi}{n}\ldots\sin\frac{n-1}{n}\pi.$$

2.13 To Find the Values of the Integrals

(i) $\int_0^\infty e^{-ax}\cos bx.\,x^{m-1}dx$

and (ii) $\int_0^\infty e^{-ax}\sin bx.\,x^{m-1}dx$.

We have

$$\int_0^\infty e^{-ax}e^{ibx}x^{m-1}dx = \int_0^\infty e^{-(a-ib)x}x^{m-1}dx = \frac{\Gamma(m)}{(a-ib)^m}$$

$$= (a-ib)^{-m}\,\Gamma(m). \qquad \ldots(1)$$

Let us first separate $(a-ib)^{-m}$ into real and imaginary parts.

Put $a = k\cos\alpha$ and $b = k\sin\alpha$ so that

$\alpha = \tan^{-1}(b/a)$ and $k = \sqrt{(a^2+b^2)}$.

Then $(a-ib)^{-m} = [k(\cos\alpha - i\sin\alpha)]^{-m}$

$= k^{-m}(\cos\alpha - i\sin\alpha)^{-m}$

$= k^{-m}(\cos m\alpha + i\sin m\alpha)$, by De-Moivre's theorem.

Now from (1), we have

$$\int_0^\infty e^{-ax}e^{ibx}x^{m-1}dx = k^{-m}(\cos m\alpha + i\sin m\alpha)\Gamma(m)$$

or $\int_0^\infty e^{-ax}(\cos bx + i\sin bx)x^{m-1}dx$

$$= \frac{\Gamma(m)}{k^m}(\cos m\alpha + i\sin m\alpha),$$

[$\because e^{i\theta} = \cos\theta + i\sin\theta$, by Euler's theorem]

or $\int_0^\infty e^{ax}\cos bx.\,x^{m-1}dx + i\int_0^\infty e^{-ax}\sin bx.\,x^{m-1}dx$

$$= \frac{\Gamma(m)}{k^m}\cos m\alpha + i\frac{\Gamma(m)}{k^m}\cos m\alpha. \qquad \ldots(2)$$

Equating real and imaginary parts in (2), we get

$$\int_0^{\infty} e^{-ax}\cos bx.\, x^{m-1}dx = \frac{\Gamma(m)\cos m\alpha}{k^m},$$

and $$\int_0^{\infty} e^{-ax}\sin bx.\, x^{m-1}dx = \frac{\Gamma(m)}{k^m}\sin m\alpha,$$

where $k = \sqrt{(a^2+b^2)}$ and $\alpha = \tan^{-1}(b/a)$.

Deductions

(i) If we put $a = 0$, then $\alpha = \pi/2$ and $k = b$.

Hence $$\int_0^{\infty} x^{m-1}\cos bx dx = \frac{\Gamma(m)}{b^m}\cos\frac{m\pi}{2}$$

and $$\int_0^{\infty} x^{m-1}\sin bx\, dx = \frac{\Gamma(m)}{b^m}\sin\frac{m\pi}{2}.$$

(ii) If we put $m = 1$, then

$$\int_0^{\infty} e^{-ax}\cos bx\, dx = \frac{\Gamma(1)}{k}\cos\alpha = \frac{k\cos\alpha}{k^2} = \frac{a}{a^2+b^2}$$

and $$\int_0^{\infty} e^{-ax}\sin bx\, dx = \frac{\Gamma(1)}{k}\sin\alpha = \frac{k\sin\alpha}{k^2} = \frac{b}{a^2+b^2}.$$

2.14 Dirichlet's Theorem for Three Variables

If l, m, n are all positive, then the triple integral

$$\iiint x^{l-1}y^{m-1}z^{n-1}dx\, dy\, dz = \frac{\Gamma(l)\Gamma(m)\Gamma(n)}{\Gamma(l+m+n+1)},$$

where the integral is extended to all positive values of the variables x, y and z subject to the condition $x + y + z \le 1$.

Proof:

Let us first consider the double integral

$$I_2 = \iint x^{l-1}y^{m-1}dx\, dy,$$

where the integral is extended to all positive values of the variables x and y subject to the condition $x + y \le 1$.

Obviously the region of integration of I_2, in the 2 dimensional Euclidean space, is bounded by the straight lines $x = 0$, $y = 0$ and $x + y = 1$. The limits of integration for this region can be expressed as $0 \le x \le 1$, $0 \le y \le 1 - x$.

$$\therefore I_2 = \int_{x=0}^{1}\int_{y=0}^{1-x} x^{l-1}y^{m-1}dx\, dy$$

$$= \int_0^1 x^{l-1}\left[\frac{y^m}{m}\right]_0^{1-x} dx = \int_0^1 \frac{1}{m} d^{l-1}(1-x)^m dx$$

$$= \frac{1}{m}\int_0^1 x^{l-1}(1-x)^{m+1-1} dx = \frac{1}{m} B(l, m+1),$$

by the def. of Beta function

$$= \frac{1}{m}\frac{\Gamma(l)\Gamma(m+1)}{\Gamma(l+m+1)} = \frac{1}{m}\frac{\Gamma(l).m\Gamma(m)}{\Gamma(l+m+1)}, \qquad [\because \Gamma(n+1) = n\Gamma(n)]$$

$$= \frac{1}{m}\frac{\Gamma(l)\Gamma(m)}{\Gamma(l+m+1)}. \qquad ...(1)$$

This is Dirichlet's theorem for two variables.

Now consider the double integral

$$U_2 = \iint x^{l-1} y^{m-1} dx\, dy,$$

where the integral is extended to all positive values to the variables x and y subject to the condition $x + y \le h$.

We have $x + y \le h \Rightarrow \frac{x}{h} + \frac{y}{h} \le 1$.

So pitting $x/h = u$ and $y/h = v$ so that $dx = h\,du$ and $dy = h\,dv$, the integral U_2 becomes

$$U_2 = \iint (hu)^{l-1}(hv)^{m-1} h^2 du\, dv$$

$$= h^{l+m}\iint u^{l-1} v^{m-1} du\, dv, \text{ where } u + v \le 1$$

$$= h^{l+m}\frac{\Gamma(l)\Gamma(m)}{\Gamma(l+m+1)}, \text{dy (1)}. \qquad ...(2)$$

Now we consider the triple integral

$$I_3 = \iiint x^{l-1} y^{m-1} z^{n-1} dx\, dy\, dz,$$

subject to the condition $x + y + z \le 1$ *i.e.*, $y + z \le 1 - x$ and $0 \le x \le 1$.

We have

$$I_3 = \int_{x=0}^1 \left[\iint y^{m-1} z^{n-1} dy\, dz\right] x^{l-1} dx, \text{ where } y + z \le 1 - x$$

$$= \int_0^1 (1-x)^{m+n}\frac{\Gamma(m)\Gamma(n)}{\Gamma(m+n+1)} x^{l-1} dx, \qquad \text{by using (2)}$$

$$= \frac{\Gamma(m)\Gamma(n)}{\Gamma(m+n+1)}\int_0^1 x^{l-1}(1-x)^{m+n+1-1} dx$$

$$= \frac{\Gamma(m)\Gamma(n)}{\Gamma(m+n+1)} B(l, m+n+1)$$

$$= \frac{\Gamma(m)\Gamma(n)}{\Gamma(m+n+1)} \cdot \frac{\Gamma(l)\Gamma(m+n+1)}{\Gamma(l+m+n+1)}$$

$$= \frac{\Gamma(l)\Gamma(m)\Gamma(n)}{\Gamma(l+m+n+1)},$$ which proves the required result.

Remark:

The triple integral

$$\iiint x^{l-1} y^{m-1} z^{n-1} dx\, dy\, dz = h^{l+m+n} \frac{\Gamma(l)\Gamma(m)\Gamma(n)}{\Gamma(l+m+n+1)}.$$

where the integral is extended to all positive of the variables x, y and z subject to the condition $x + y + z \le h$.

Alternative proof of Dirichlet's theorem for three variables.

Let $I_3 = \iiint x^{l-1} y^{m-1} z^{n-1} dx\, dy\, dz,$

where the integral is extended to all positive values of the variables x, y and z subject to the condition $x + y + z \le 1$.

Obviously the region of integration, in the 3-dimensional Euclidean space, is the volume bounded by the coordinate planes $x = 0$, $y = 0$, $z = 0$ and the plane $x + y + z = 1$. After a little geometric consideration, we observe that the limits of integration for this region can be expressed as

$$0 \le x \le 1,\ 0 \le y \le 1 - x\ 0 \le z \le 1 - x - y.$$

Hence the triple integral I_3 may be written as

$$I_3 = \int_{x=0}^{1} \int_{y=0}^{1-x} \int_{z=0}^{1-x-y} x^{l-1} y^{m-1} z^{n-1} dx\, dy\, dz$$

$$= \int_0^1 \int_0^{1-x} x^{l-1} y^{m-1} \left[\frac{z^n}{n}\right]_0^{1-x-y} dx\, dy$$

$$= \frac{1}{2} \int_0^1 \int_0^{1-x} x^{l-1} y^{m-1} (1-x-y)^n dx\, dy$$

$$= \frac{1}{n} \int_0^1 x^{l-1} \left[\int_0^{1-x} y^{m-1} \{(1-x)-y\}^n dy\right] dx.$$

To integrate w.r.t. y, put $y = (1 - x)\, t$ so that $dy = (1 - x)\, dt$; also when $y = 0$, $t = 0$ and when $y = 1 - x$, $t = 1$.

∴ the required integral

$$I_3 = \frac{1}{n}\int_0^1 x^{l-1}\left[\int_0^1 (1-x)^{m-1} t^{m-1}\{(1-x)^n(1-t)^n\}(1-x)dt\right]dx$$

$$= \frac{1}{n}\int_0^1\int_0^1 x^{l-1}(1-x)^{m+n} t^{m-1}(1-t)^n dx\, dt$$

$$= \frac{1}{n}\int_0^1 x^{l-1}(1-x)^{m+n} dx \times \int_0^1 t^{m-1}.(1-t)^n dt$$

$$= \frac{1}{n}B(l, m+n+1)B(m, n+1), \qquad \text{(by the definition of Beta function)}$$

$$= \frac{1}{n}\cdot\frac{\Gamma(l)\Gamma(m+n+1)}{\Gamma(l+m+n+1)}\cdot\frac{\Gamma(m)\Gamma(n+1)}{\Gamma(m+n+1)} = \frac{\Gamma(l)\Gamma(m)\Gamma(n)}{\Gamma(l+m+n+1)},$$

$[\because \Gamma(n+1) = n\,\Gamma(n)]$.

Note: Dirichlet's theorem holds good even if the condition is taken as $x + y + z < 1$ in place of $x + y + z \leq 1$.

Cor. : *Evaluate without using Dirichlet's theorem*

$$\iiint x^p y^q z^r dx\, dy\, dz,$$

where x, y, z are always positive and $x + y + z \leq 1$.

2.15 Dirichlet's Theorem for n Variables

The theorem states that

$$\iint\ldots\int x_1^{l_1-1}\ldots x_2^{l_2-1}\ldots x_n^{l_n-1}\, dx_1 dx_2\ldots dx_n$$

$$= \frac{\Gamma(l_1)\Gamma(l_2)\ldots\Gamma(l_n)}{\Gamma(1+l_1+l_2+\ldots+l_n)},$$

where the integral is extended to all positive values of the variables $x_1, x_2, \ldots, x_n$ *subject to the condition* $x_1 + x_2 + \ldots + x_n \leq 1$.

Proof:

We shall prove the theorem by mathematical induction.

To start the induction we shall show that the theorem is true for two variables *i.e.*, for n = 2.

So let us consider the integral

$$I_2 = \iint x_1^{l_1-1} x_2^{l_2-1} dx_1 dx_2$$

subject to the condition $x_1 + x_2 \leq 1$.

$$I_2 = \frac{\Gamma(l_1)\Gamma(l_2)}{\Gamma(1+l_1+l_2)}.$$

The result (1) shows that the theorem is true for two variables *i.e.*, for n = 2.

Now assume as our induction hypothesis that the theorem is true for n variables *i.e.*, assume that

$$In = \iint \dots \int x_1^{l_1-1} x_2^{l_2-1} \dots x_n^{l_n-1}\, dx_1 dx_2 \dots dx_n$$

$$= \frac{\Gamma(l_1)\Gamma(l_2)\dots\Gamma(l_n)}{\Gamma(1+l_1+l_2+\dots+l_n)}, \qquad \dots(2)$$

subject to the condition $x_1 + x_2 + \dots + x_n \le 1$.

If the condition be $x_1 + x_2 + \dots + x_n \le h$, then putting

$\frac{x_1}{h} = u_1$, $\frac{x_2}{h} = u_2$, ..., $\frac{x_n}{h} = u_n$, so that

$dx_1 = hdu_1$, $dx_2 = hdu_2$, ..., $dx_n = hdu_n$, we have

$$\iint \dots \int x_1^{l_1-1} x_2^{l_2-1} \dots x_n^{l_n-1} dx_1 dx_2 \dots dx^n$$

$$= h^{l_1+l_2+\dots+l_n} \iint \dots \int u_1^{l_1-1} u_2^{l_2-1} \dots u_n^{l_n-1} du_1 du_2 \dots du_n$$

subject to the condition $u_1 + u_2 + \dots + u_n \le 1$

$$= h^{l_1+l_2+\dots+l_n} \frac{\Gamma(l_1)\Gamma(l_2)\dots\Gamma(l_n)}{\Gamma(1+l_1+l_2+\dots+l_n)} \quad \text{using the assumed result (2)} \quad \dots(3)$$

Now for n + 1 variables the condition is

$x_1 + x_2 + \dots + x_n + x_{n+1} \le 1$

i.e., $x_2 + x_3 + \dots + x_n + x_{n+1} \le 1 - x_1$, and $0 \le x_1 \le 1$.

We then have

$$\iint \dots \int x_1^{l_1-1} x_2^{l_2-1} \dots x_{n+1}^{l_{n+1}-1} dx_1 dx_2$$

$\dots dx_n\, dx_{n+1}$, where $x_1 + x_2 + \dots + x_{n+1} \le 1$

$$= \int_{x_1=0}^{1} x_1^{l_1-1} \left[\iint \dots \int x_2^{l_2-1} \dots x_{n+1}^{l_{n+1}} dx_2 \dots dx_{n+1} \right] dx_1$$

$$= \int_{x_1=0}^{1} x_1^{l_1-1} \cdot \frac{\Gamma(l_2)\Gamma(l_3)\dots\Gamma(l_{n+1})}{\Gamma(1+l_2+l_3+\dots+l_n+l_{n+1})}.$$

$(1-x_1)l_2 + l_3 + \dots + l_{n+1}\, dx_1$, using (3)

$$= \int_{x_1=0}^{1} x_1^{l_1-1} \cdot \frac{\Gamma(l_2)\Gamma(l_3)\ldots\Gamma(l_{n+1})}{\Gamma(1+l_2+\ldots+l_n+l_{n+1})}$$

$$\int_0^1 x_1^{l_1-1}(1-x_1)^{(1+l_2+l_3+\ldots+l_{n+1})-1}dx_1$$

$$= \frac{\Gamma(l_2)\Gamma(l_3)\ldots\Gamma(l_{n+1})}{\Gamma(1+l_2+\ldots+l_n+l_{n+1})} \cdot \frac{\Gamma(l_1)\Gamma(1+l_2+\ldots+l_{n+1})}{\Gamma(1+l_1+l_2+\ldots+l_n+l_{n+1})}$$

$$= \frac{\Gamma(l_1)\Gamma(l_2)\ldots\Gamma(l_{n+1})}{\Gamma(1+l_1+l_2+\ldots+l_{n+1})}. \qquad \ldots(4)$$

The result (4) shows that the theorem holds for (n + 1) variables if it holds for n variables. But we have seen that the theorem is true for two variables. Hence by mathematical induction the theorem is true for all values fo n.

2.16 Liouville's Extension of Dirichlet's Theorem

If the variables x, y, z are all positive such that

$h_1 \le x + y + z \le h_2$, then the triple integral

$$\iiint f(x+y+z)x^{l-1}y^{m-1}z^{n-1}dx\,dy\,dz$$

$$= \frac{\Gamma(l)\Gamma(m)\Gamma(n)}{\Gamma(1+m+n)}\int_{h_1}^{h_2} f(u)u^{l+m+n-1}du.$$

Proof:

Let $I = \iiint d^{l-1}y^{m-1}z^{n-1}dx\,fy\,dz$, (integrated over some region)

Subject to the condition x + y + z ≤ u, we have by Dirichlet's theorem

$$I = u^{l+m+n}\frac{\Gamma(l)\Gamma(m)\Gamma(n)}{\Gamma(l+m+n+1)}. \qquad \ldots(1)$$

If the condition be x + y + z ≤ u + *l*u, then

$$I = (u+\delta u0)^{l+m+n}\frac{\Gamma(l)\Gamma(m)\Gamma(n)}{\Gamma(l+m+n+1)}. \qquad \ldots(2)$$

Therefore the value of the integral I extended to all such positive values of the variables as make the sum of the variables lie between u and u + du is

$$= \frac{\Gamma(l)\Gamma(m)\Gamma(n)}{\Gamma(l+m+n+1)}[(u+\delta u)l^{l+m+n} - u^{l+m+b}], \text{ [subtracting (2) from (1)]}$$

$$= \frac{\Gamma(l)\Gamma(m)\Gamma(n)}{\Gamma(l+m+n+1)}u^{l+m+n}\left[\left(1+\frac{\delta u}{u}\right)^{l+m+n} - 1\right]$$

$$= \frac{\Gamma(l)\Gamma(m)\Gamma(n)}{\Gamma(l+m+n+1)} u^{l+m+n} \left[1 + (l+m+n)\frac{\delta u}{u} + \ldots - 1 \right]$$

$$= \frac{\Gamma(l)\Gamma(m)\Gamma(n)}{\Gamma(l+m+n+1)} (l+m+n) u^{l+m+n-1} \delta u,$$

to the first order of approximation

$$= \frac{\Gamma(l)\Gamma(m)\Gamma(n)}{\Gamma(l+m+n)} u^{l+m+n-1} \delta u.$$

Now consider the integral

$$\iiint f(x+y+z) x^{l-1} y^{m-1} z^{n-1} dx\, dy\, dz,$$

subject tot he condition $h_1 \le x + y + z \le h_2$.

If $x + y + z$ lies between u and $u + \delta u$, the value of $f(x + y + z)$ can only differ from $f(u)$ by a small quantity of the same order as δu. Hence neglecting square of du, the part of the integral

$$\iiint f(x+y+z) x^{l-1} y^{m-1} z^{n-1} dx\, dy\, dz$$

which arises from supposing the sum of the variables to lie between u and $u + \delta u$ is ultimately equal to

$$\frac{\Gamma(l)\Gamma(m)\Gamma(n)}{\Gamma(l+m+n)} f(u) . u^{l+m+n-1} \delta u.$$

Therefore the whole integral $\iiint f(x+y+z) x^{l-1} y^{m-1} z^{n-1} dx\, dy\, dz$

where $h_1 \le x + y + z \le h_2$, is equal to

$$\frac{\Gamma(l)\Gamma(m)\Gamma(n)}{\Gamma(l+m+n)} \int_{h_1}^{h_2} f(u) . u^{l+m+n-1} du.$$

Remark:

The above theorem holds good even if we take the condition as

$h_1 < x + y < z < h_2$ in place of $h_1 \le x + y + z \le h_2$.

MISCELLANEOUS EXAMPLES

Example 1:

Evaluate the integral $\int_a^b (x-a)^p (b-x)^q\, dx$, *where p and q are positive integers.*

Solution:

Let $I = \int_a^b (x-a)^p (b-x)^q\, dx.$

Put $x = a\cos^2\theta + b\sin^2\theta$ so that

$$dx = -2a\cos\theta\sin\theta\,d\theta + 2b\sin\theta\cos\theta\,d\theta$$

i.e., $dx = 2(b-a)\cos\theta\sin\theta\,d\theta.$

Also $x - a = a\cos^2\theta + b\sin^2\theta - a = b\sin^2\theta - a(1-\cos^2\theta)$

$= b\sin^2\theta - a\sin^2\theta = (b-a)\sin^2\theta$

and $b - x = b - a\cos^2\theta - b\sin^2\theta = b(1-\sin^2\theta) - a\cos^2\theta$

$= (b-a)\cos^2\theta.$

To find the limits for θ, when $x = a$, we have

$a = a\cos^2\theta + b\sin^2\theta$

i.e., $(b-a)\sin^2\theta = 0$ *i.e.,* $\sin^2\theta = 0$ as $a \neq b$

i.e., $\theta = 0$

and when $x = b$,

we have $b = a\cos^2\theta + b\sin^2\theta$

i.e., $(a-b)\cos^2\theta = 0$

i.e., $\cos^2\theta = 0$ as $a \neq b$

ie., $\theta = \pi/2.$

Thus the new limits for θ are 0 to $\pi/2$. Hence the given integral

$$I = \int_0^{\pi/2} (b-a)^p \sin^{2p}\theta.(b-a)^q \cos^{2q}\theta.2(b-a)\cos\theta\sin\theta\,d\theta$$

$$= 2(b-a)^{p+q+1}\int_0^{\pi/2} \sin^{2p+1}\theta\cos^{2q+1}\theta\,d\theta$$

$$= 2(b-a)^{p+q+1}\frac{\Gamma\left\{\frac{1}{2}(2p+1+1)\right\}\Gamma\left\{\frac{1}{2}(2q+1+1)\right\}}{2\Gamma(2p+1+2q+1+2)},$$

provided $2p + 1 > -1$ and $2q + 1 > -1$

i.e., $p > -1$ and $q > -1$ which is so because

p and q are given to be +ive integers

$$= (b-a)^{p+q+1}\frac{\Gamma(p+1)\Gamma(q+1)}{\Gamma(p+q+1+1)}$$

$$= (b-a)^{p+q+1}\frac{p!q!}{(p+q+1)!},$$

because $\Gamma(n+1) = n!$

if n is a positive integer.

Example 2:

Prove that $B(m,m).\ B\left(m+\frac{1}{2}, m+\frac{1}{2}\right) = \frac{\pi m^{-1}}{2^{4m-1}}$.

Solution:

We have $B(m, m)\ .\ B\left(m+\frac{1}{2}, m+\frac{1}{2}\right)$

$$= \frac{\Gamma(m).\Gamma(m)}{\Gamma(m+m)} \cdot \frac{\Gamma\left(m+\frac{1}{2}\right)\Gamma\left(m+\frac{1}{2}\right)}{\Gamma\left(m+\frac{1}{2}+m+\frac{1}{2}\right)}, \quad \left[\because B(m,n) = \frac{\Gamma(m)\Gamma(n)}{\Gamma(m+n)}\right]$$

$$= \frac{\left[\Gamma(m)\Gamma\left(m+\frac{1}{2}\right)\right]^2}{\Gamma(2m).\Gamma(2m+1)} = \frac{\left[\Gamma(m)\Gamma\left(m+\frac{1}{2}\right)\right]^2}{\Gamma(2m).2m\Gamma(2m)},$$

$[\because \Gamma(p+1) = p\,\Gamma(p)]$

$$= \frac{1}{2m}\left[\frac{\Gamma(m)\Gamma\left(m+\frac{1}{2}\right)}{\Gamma(2m)}\right]^2 = \frac{1}{2m} \cdot \left[\frac{\sqrt{\pi}}{2^{2m-1}}\right]^2$$

$$\left[\because \text{by Dupliaction formula, (11 page 275)}, \frac{\Gamma(m)\Gamma\left(m+\frac{1}{2}\right)}{\Gamma(2m)} = \frac{\sqrt{\pi}}{2^{2m-1}}\right]$$

$$= \frac{1}{2m} \cdot \frac{\pi}{2^{4m-2}} = \frac{\pi m^{-1}}{2^{4m-1}}.$$

Example 3:

Find the value of $\Gamma\left(\frac{1}{9}\right)\Gamma\left(\frac{2}{9}\right)\Gamma\left(\frac{3}{9}\right)\ldots\Gamma\left(\frac{8}{9}\right)$.

Solution:

We know that

$$\Gamma\left(\frac{1}{n}\right)\Gamma\left(\frac{2}{n}\right)\Gamma\left(\frac{3}{n}\right)\ldots\Gamma\left(\frac{n-1}{n}\right) = \frac{(2\pi)^{(n-1)/2}}{n^{1/2}},$$

where n is a positive integer.

Putting n = 9 in the above relation, we get

$$\Gamma\left(\frac{1}{9}\right)\Gamma\left(\frac{2}{9}\right)\ldots\Gamma\left(\frac{8}{9}\right)=\frac{(2\pi)^{(n-1)/2}}{9^{1/2}}=\frac{(2\pi)^4}{3}$$

$= (16/3)\,\pi^4$

Example 3:

Show that **B** (m, n) = **B** (m + 1, n) + **B** (m, n + 1), for m > 0, n > 0.

Solution:

We know that $\mathbf{B}(m, n)=\dfrac{\Gamma(m)\Gamma(n)}{\Gamma(m+n)}$.

$$\therefore\ \mathbf{B}(m+1, n)=\frac{\Gamma(m+1)\Gamma(n)}{\Gamma(m+1+n)}$$

and $\qquad \mathbf{B}(m, n+1)=\dfrac{\Gamma(m)\Gamma(n+1)}{\Gamma(m+n+1)}$.

adding, we get **B** (m + 1, n) + **B** (m, n + 1)

$$=\frac{\Gamma(m+1)\Gamma(n)+\Gamma(m)\Gamma(n+1)}{\Gamma(m+n+1)}$$

$$=\frac{m\Gamma(m)\Gamma(n)+\Gamma(m).n(n)}{(m+n)\Gamma(m+n)}$$

$$=\frac{(m+n)\Gamma(m)\Gamma(n)}{(m+n)\Gamma(m+n)}=\frac{\Gamma(m)\Gamma(n)}{\Gamma(m+n)}=\mathbf{B}(m, n),$$

provided m > 0 and n > 0.

Example 4:

Show that $\Gamma(0\cdot 1)\,\Gamma(0\cdot 2)\,G(0\cdot 3)\ldots\Gamma(0\cdot 9)=\dfrac{(2\pi)^{9/2}}{\sqrt{10}}$.

Solution:

Do your self.

Put n = 10.

Example 5:

Show that

(i) $2^n\Gamma\left(n+\dfrac{1}{2}\right)=1.3.5\ldots.(2n-1)\sqrt{\pi}$, where n is a +ive integer,

(ii) $\Gamma\left(\dfrac{3}{2}-x\right)\Gamma\left(\dfrac{3}{2}+x\right)=\left(\dfrac{1}{4}-x^2\right)\pi\sec\pi x$, $\quad$ *(provided* $-1<2x<1$*)*

Solution:

(i) We have

$$\Gamma\left(n+\frac{1}{2}\right)=\Gamma\left(n-\frac{1}{2}\right)\Gamma\left(n-\frac{1}{2}\right)$$

$$=\left(n-\frac{1}{2}\right)\Gamma\left(n-\frac{3}{2}\right)\Gamma\left(n-\frac{3}{2}\right)$$

$$\left(n-\frac{1}{2}\right)\left(n-\frac{3}{2}\right)\left(n-\frac{5}{2}\right)\ldots\frac{3}{2}\cdot\frac{1}{2}\cdot\Gamma\left(\frac{1}{2}\right)$$

$$\frac{2n-1}{2}\cdot\frac{2n-3}{2}\cdot\frac{2n-5}{2}\ldots\frac{3}{2}\cdot\frac{1}{2}\cdot\sqrt{\pi}$$

$$\frac{1}{2^n}(1n-1)(2n-3)(2n-5)\ldots3.1.\sqrt{\pi}$$

$$\therefore\ 2n\Gamma\left(n+\frac{1}{2}\right)=1.3.5.(2n-1)\sqrt{\pi}.$$

(ii) We have $\Gamma\left(\frac{3}{2}-x\right)\Gamma\left(\frac{3}{2}+x\right)$

$$=\left(\frac{1}{2}-x\right)\Gamma\left(\frac{1}{2}-x\right).\left(\frac{1}{2}+x\right)\Gamma\left(\frac{1}{2}+x\right)$$

$$=\left(\frac{1}{4}-x^2\right)\Gamma\left(\frac{1-2x}{2}\right)\Gamma\left(\frac{1+2x}{2}\right)$$

$$=\left(\frac{1}{4}-x^2\right)\Gamma\left(\frac{1-2x}{2}\right)\Gamma\left(1-\frac{1+2x}{2}\right)$$

$$=\left(\frac{1}{4}-x^2\right)\frac{\pi}{\sin\left(\frac{1-2x}{2}\pi\right)}=\left(\frac{1}{4}-x^2\right)\frac{\pi}{\sin\left(\frac{1}{2}\pi-x\pi\right)}$$

$$=\left(\frac{1}{4}-x^2\right).\frac{\pi}{\cos x\pi}=\left(\frac{1}{4}-x^2\right)\cdot\pi\sec\pi x$$

Example 6:

With certain restrictions on the values of a, b, m and n, prove that

$$\int_0^\infty\int_0^\infty e^{-(ax^2+by^2)}x^{2m-1}dx\,dy=\frac{\Gamma(m)\Gamma(n)}{4a^mb^n}.$$

Solution:

Let us denote the given integral by I. Then

$$I = \int_0^\infty e^{-ax^2} x^{2m-1} dx \times \int_0^\infty e^{-by^2} y^{2n-1} dy = I_1 \times I_2.$$

To evaluate I_1, put $ax^2 = t$ so that $2ax\,dx = dt$.

$$\therefore I_1 = \int_0^\infty e^{-t} (t/a)^{(2m-1)/2} \cdot \frac{dt}{2\sqrt{(at)}}$$

$$= \frac{1}{2a^m} \int_0^\infty e^{-t} t^{m-1} dt = \frac{\Gamma(m)}{2a^m}, \text{ provided a and m are +ive.}$$

Similarly, $I_2 = \frac{\Gamma(n)}{2b^n}$, provided b and n are +ive.

Hence $I = \frac{\Gamma m \Gamma n}{4a^m b^n}$.

Example 7:

Show that the sum of the series

$$\frac{1}{x+1} + m\frac{1}{n+2} + \frac{m(m+1)}{2!} \cdot \frac{1}{n+3} + \frac{m(m+1)(m+2)}{3!} \cdot \frac{1}{n+4} + \ldots$$

is $\frac{\Gamma(n+1)\Gamma(1-m)}{\Gamma(n-m+2)}$, where $-1 < n < 1$.

Solution:

We have $\frac{\Gamma(n+1)\Gamma(1-m)}{\Gamma(n-m+2)} = B(+1, 1-m)$

$$= \int_0^1 x^n (1-x)^{-m} dx$$

$$= \int_0^1 x^n \left[1 + mx + \frac{m(m+1)}{2!} x^2 + \frac{m(m+1)(m+2)}{3!} x^3 + \ldots\right] dx$$

$$= \int_0^1 \left[x^n + mx^{n+1} + \frac{m(m+1)}{2!} x^{n+2} + \frac{m(m+1)(m+2)}{3!} n^{n+3} + \ldots\right] dx$$

$$= \left[\frac{x^{n+1}}{n+1} + \frac{x^{n+2}}{n+2} + \frac{m(m+1)}{2!}\frac{x^{n+3}}{n+3} + \frac{m(m+1)(m+2)}{3!}\frac{x^{n+4}}{+4} + \ldots\right]_0^1$$

$$= \frac{1}{n+1} + m.\frac{1}{n+2} + \frac{m(m+1)}{2!} \cdot \frac{1}{n+3} + \frac{m(m+1)(m+2)}{3!}\frac{1}{n+4} + \ldots$$

Example 8:

Prove that $\int_0^\infty \frac{\sin bz}{z} dz = \frac{\pi}{2}$.

Solution:

We have

$$I = \int_0^\infty \int_0^\infty e^{-xz} \sin bz\, dx\, dz$$

$$= \int_0^\infty \left[\frac{e^{-xz}}{-z}\right]_0^\infty \sin bz dz, \text{ on first integrating w.r.t. x}$$

$$= \int_0^\infty \frac{\sin bz}{z} dz. \qquad ...(1)$$

Again on first integrating w.r.t. z, we have

$$I = \int_0^\infty \int_0^\infty e^{-xz} \sin bz\, dx\, dz = \int_0^\infty \left[\int_0^\infty e^{-xz} \sin bz\, dz\right] dx$$

$$= \int_0^\infty \frac{b}{b^2 + x^2} dx,$$

$$= \left[\tan^{-1} \frac{x}{b}\right]_0^\infty = \frac{\pi}{2}. \qquad ...(2)$$

Hence equating the two values (1) and (2) of I, we have

$$\int_0^\infty \frac{\sin bz}{z} dx = \frac{\pi}{2}.$$

Example 9:

Show that $\int_0^\infty \cos(bz^{1/z})dz = \frac{1}{b^2}\Gamma(n+1).\cos\frac{n\pi}{2}.$

Solution:

Put $z^{1/n} = x$ *i.e.*, $z = x^n$,

so that $dz = nx^{n-1}\, dx$.

$$\therefore \int_0^\infty \cos(bz^{1/z})dz = \int_0^\infty \cos(bx). nx^{n-1} dx$$

$$= n\int_0^\infty x^{n-1} \cos(bx) dx$$

$$= \text{real part of } n\int_0^\infty e^{-ibx} x^{n-1} dx$$

$$= \text{real part of } n\frac{\Gamma(n)}{(ib)^n}$$

$$= \text{real part of } \frac{n\Gamma(n)}{b^n} \cdot \left(\cos\frac{1}{2}\pi + i\sin\frac{1}{2}\pi\right)^{-n}$$

$$= \text{real part of } \frac{\Gamma(n+1)}{b^n}\left(\cos\frac{n\pi}{2} - i\sin\frac{n\pi}{2}\right)$$

$$= \frac{1}{b^n}\cdot\Gamma(n+1).\cos\left(\frac{n\pi}{2}\right).$$

Example 10:

Prove that

(i) $\int_0^2 (8-x^3)^{-1/3}\,dx = \frac{2\pi}{3\sqrt{3}}$

(ii) $\int_0^1 x^m(1-x^n)^p\,dx = \frac{1}{n}B\left(\frac{m+1}{n}, p+1\right).$

(iii) $\int_0^1 (1-x^n)^{1/n}\,dx = \frac{1}{n}\frac{\left[\Gamma\left(\frac{1}{n}\right)\right]^2}{2\Gamma\left(\frac{2}{n}\right)}.$

(iv) $\int_0^1 x^{m-1}(1-x^2)^{n-1}\,dx = \frac{1}{2}B\left(\frac{m}{2}, n\right).$

Solution:

(i) Let

$$I = \int_0^2 (8-x^3)^{-1/3}\,dx = \int_0^2 x^{-2}(8-x^3)^{-1/3}x^2dx.$$

Put $x^3 = 8y$ so that $3x^2\,dx = 8\,dy$.

When $x = 0$, $y = 0$

and when $x = 2$, $y = 1$.

$$\therefore I = \int_0^1 8^{-2/3}y^{-2/3}(8-8y)^{-1/3}.\frac{8}{3}dy$$

$$= \frac{1}{3}\int_0^1 y^{-2/3}(1-y)^{-1/3}\,dy$$

$$= \frac{1}{3}\int_0^1 y^{(1/3)-1}(1-y)^{(2/3)-1}\,dy = \frac{1}{3}B\left(\frac{1}{3},\frac{2}{3}\right)$$

$$= \frac{1}{3}\frac{\Gamma\left(\frac{1}{3}\right)\Gamma\left(\frac{2}{3}\right)}{\Gamma\left(\frac{1}{3}+\frac{2}{3}\right)} = \frac{1}{3}\frac{\Gamma\left(\frac{1}{3}\right)\Gamma\left(1-\frac{1}{3}\right)}{\Gamma 1} = \frac{1}{3}\cdot\frac{\pi}{\sin\frac{1}{3}\pi}$$

$$= \frac{\pi}{3} \cdot \frac{2}{\sqrt{3}} = \frac{2\pi}{3\sqrt{3}}.$$

(ii) Let $I = \int_0^1 x^m (1 - x^n)^p \, dx.$

Put $x^n = y$ or $x = y^{1/n}$.

Then $dx = \frac{1}{n} y^{(1/n)-1} dy.$

When $x = 0$, $y = 0$

and when $x = 1$, $y = 1$.

$$\therefore I = \int_0^1 (y^{1/n})^m (1 - y)^p \cdot \frac{1}{n} y^{(1/n)-1} dy$$

$$= \frac{1}{n} \int_0^1 y^{(m/n)+(1/n)-1} (1 - y)^{(p+1)-1} dy$$

$$= \frac{1}{n} \int_0^1 y^{\{(m+1)/n\}-1} (1 - y)^{(p+1)-1} dy$$

$$= \frac{1}{n} B\left(\frac{m+1}{n}, p+1\right).$$

(iii) Let $I = \int_0^1 (1 - x^n)^{1/n} dx.$

Put $x^n = y$ or $x = y^{1/n}$.

Then $dx = (1/n)\, y^{(1/n)-1}\, dy$.

When $x = 0$, $y = 0$

and when $x = 1$, $y = 1$.

$$\therefore I = \int_0^1 (1 - y)^{1/n} \cdot \frac{1}{n} y\left(\frac{1}{n}\right) - 1$$

$$= \frac{1}{n} \int_0^1 y^{(1/n)-1} (1 - y)^{\{(1/n)+1\}-1} dy$$

$$= \frac{1}{n} B\left(\frac{1}{n}, \frac{1}{n} + 1\right) = \frac{1}{n} \frac{\Gamma(1/n)\Gamma\{(1/n)+1\}}{\Gamma\{(2/n)+1\}}$$

$$= \frac{1}{n} \cdot \frac{\Gamma(1/n).(1/n)\Gamma(1/n)}{(2/n).\Gamma(2/n)} = \frac{1}{n} \frac{[\Gamma(1/n)]^2}{2\Gamma(2/n)}.$$

(iv) Let

$$I = \int_0^1 x^{m-1} (1 - x^2)^{n-1} dx = \int_0^1 x^{m-2} (1 - x^2)^{n-1} x \, dx.$$

Put $x^2 = y$

so that $2x\,dx = dy$.

When $x = 0$, $y = 0$

and when $x = 1$, $y = 1$.

$$\therefore I = \int_0^1 y^{(m-2)/2}(1-y)^{n-1} \cdot \frac{1}{2} dy$$

$$= \frac{1}{2}\int_0^1 y^{(m/2)-1}(1-y)^{n-1} dy$$

$$= \frac{1}{2} B\left(\frac{m}{2}, n\right).$$

Example 11:

Show that, if m > – 1, then

$$\int_0^\infty x^m e^{-n^2x^2} dx = \frac{1}{2n^{m+1}} \Gamma\left(\frac{m+1}{2}\right).$$

Solution:

$$\text{Let } I = \int_0^\infty x^m e^{-n^2x^2} dx = \int_0^\infty x^{m-1} e^{-n^2x^2} x\,dx.$$

Put $n^2 x^2 = t$,

so that $2n^2 x\,dx = dt$.

Also $x = t^{1/2}/n$.

When $x = 0$, $t = 0$

and when $x \to \infty$, $t \to \infty$.

$$\therefore I = \int_0^\infty \left(\frac{t^{1/2}}{n}\right)^{m-1} e^{-1} \cdot \frac{1}{2n^2} dt$$

$$= \frac{1}{2n^{m-1} . n^2} \int_0^\infty e^{-t} t^{(m-1)/2} dt$$

$$= \frac{1}{2n^{m+1}} \int_0^\infty e^{-t} t^{\{(m+1)/2\}-1} dt$$

$$= \frac{1}{2n^{m+1}} \Gamma\left(\frac{m+1}{2}\right),$$ by def. of Gamma function, provided

$m + 1 > 0$ *i.e.* $m > -1$

Example 12:

Prove that

(i) $\int_0^1 \frac{dx}{\sqrt{(1-x^4)}} = \frac{\sqrt{2}}{8\sqrt{\pi}}\left[\Gamma\left(\frac{1}{4}\right)\right]^2$.

(ii) $\int_0^{\pi/2} \sin^p \theta \, d\theta = \frac{1}{2} B\left(\frac{1}{2}, \frac{p+1}{2}\right)$.

Solution:

(i) Let $I = \int_0^1 \frac{dx}{\sqrt{(1-x^4)}}$.

Put $x^4 = \sin^2\theta$ *i.e.*, $x = \sin^{1/2}\theta$

so that $dx = \frac{1}{2} \sin^{-1/2}\theta \cos\theta \, d\theta$.

When $x = 0, \theta = 0$

and when $x = 1, \theta = \pi/2$.

$$\therefore I = \int_0^{\pi/2} \frac{\frac{1}{2}\sin^{-1/2}\theta\cos\theta\, d\theta}{\cos\theta} = \frac{1}{2}\int_0^{\pi/2} \sin^{-1/2}\theta\cos^0\theta\, d\theta$$

$$= \frac{1}{2}\cdot\frac{\Gamma\left(\frac{1}{4}\right)\Gamma\left(\frac{1}{2}\right)}{2\Gamma\left(\frac{3}{4}\right)} = \frac{1}{4}\frac{\left[\Gamma\left(\frac{1}{4}\right)\right]^2.\sqrt{\pi}}{\Gamma\left(\frac{1}{4}\right)\Gamma\left(\frac{3}{4}\right)}$$

$$= \frac{\sqrt{\pi}}{4}\cdot\frac{\left[\Gamma\left(\frac{1}{4}\right)\right]^2}{\Gamma\left(\frac{1}{4}\right)\Gamma\left(1-\frac{1}{4}\right)} = \frac{\sqrt{\pi}}{4}\cdot\frac{\left[\Gamma\left(\frac{1}{4}\right)\right]^2}{\pi/\left(\sin\frac{1}{4}\pi\right)}$$

$$= \frac{\sqrt{\pi}}{4}\cdot\frac{\left[\Gamma\left(\frac{1}{4}\right)\right]^2}{\pi.\sqrt{2}} = \frac{\sqrt{2}}{8\sqrt{\pi}}\left[\Gamma\left(\frac{1}{4}\right)\right]^2.$$

(ii) We have $I = \int_0^{\pi/2} \sin^p\theta\, d\theta = \int_0^{\pi/2} \sin^p\theta\cos^0 d\theta$

$$= \frac{\Gamma\left(\frac{p+1}{2}\right)\Gamma\left(\frac{1}{2}\right)}{2\Gamma\left(\frac{p+2}{2}\right)} = \frac{1}{2}B\left(\frac{p+1}{2}, \frac{1}{2}\right) = \frac{1}{2}B\left(\frac{1}{2}, \frac{p+1}{2}\right).$$

Alternative Method : Put $\sin^2\theta = y$ or $\sin\theta = y^{1/2}$.

Then $\cos\theta\, d\theta = \frac{1}{2} y^{-1/2} dy$

or $d\theta = \frac{1}{2}\frac{y^{-1/2}dy}{\sqrt{(1-\sin^2\theta)}} = \frac{1}{2}(1-y)^{-1/2}y^{-1/2}dy$.

When $\theta = 0$, $y = 0$, and when $\theta = \pi/2$, $y = 1$.

$$\therefore I = \int_0^1 y^{p/2} \cdot \frac{1}{2}(1-y)^{-1/2} y^{-1/2} dy$$

$$= \frac{1}{2}\int_0^1 y^{(p-1)/2}(1-y)^{-1/2}dy$$

$$= \frac{1}{2}\int_0^1 y^{\{(p+1)/2\}-1}(1-y)^{(1/2)-1}dy$$

$$= \frac{1}{2}B\left(\frac{p+1}{2}, \frac{1}{2}\right) = \frac{1}{2}B\left(\frac{1}{2}, \frac{p+1}{2}\right).$$

Example 13:

Show that the perimeter of a loop of the curve

$$r^n = a^n \cos n\theta \text{ is } \frac{a}{n}\cdot 2^{(1/n)-1}\cdot\frac{[\Gamma(1/2b)]^2}{\Gamma(1/n)},$$

Solution:

The given curve is $r^n = a^n \cos n\theta$ ...(1)

The curve (1) is symmetrical about the initial line. We have $r = 0$ when $\cos n\theta = 0$ *i.e.*, when $n\theta = -\pi/2, \pi/2$ or $\theta = -\pi/2n, \pi/2n$. Therefore, one loop of the curve lies between the lines $\theta = -\pi/2n$ and $\theta = \pi/2n$. This loop is symmetrical about the initial line and for the portion of this loop lying above the initial line θ varies from 0 to $\pi/2n$.

Taking log of both sides of (1), we have

$n \log r = n \log a + \log \cos n\theta$.

Differentiating w.r.t. θ, we get

$$\frac{n}{r}\frac{dr}{d\theta} = 0 + \frac{1}{\cos n\theta}(-n\sin n\theta) \text{or} \frac{dr}{d\theta} = -r\tan n\theta.$$

$$\therefore \frac{ds}{d\theta} = \sqrt{\left[r^2 + \left(\frac{dr}{d\theta}\right)^2\right]} = \sqrt{(r^2 + r^2\tan^2 n\theta)} = -r\sec n\theta$$

or $ds = r\sec n\theta\, d\theta = a\cos^{1/n} n\theta \sec n\theta\, d\theta = a(\cos n\theta)(1/n)^{-1}d\theta$.

$\therefore$ The perimeter of a loop of the curve (1)

$$= 2\int_0^{\pi/2n} a(\cos n\theta)^{(1/n)-1}\, d\theta,$$

$$= 2\int_0^{\pi/2} (\cos t)^{(1/n)-1} \frac{dt}{n},$$

putting $n\theta = t$ so that $n\, d\theta = dt$

$$= \frac{2a}{n}\int_0^{\pi/2} (\cos t)^{(1/n)-1} \sin^0 t dt$$

$$= \frac{2a}{n} \frac{\Gamma(1/2n)\Gamma\left(\frac{1}{2}\right)}{2\Gamma\left\{(1/2n)+\frac{1}{2}\right\}} = \frac{a}{n} \cdot \frac{\sqrt{\pi}.[\Gamma(1/2n)]^2}{\Gamma(1/2n)\Gamma\left\{(1/2n)+\frac{1}{2}\right\}}$$

$$= \frac{a}{n} \cdot \sqrt{\pi} \frac{[\Gamma(1/2n)]^2}{\frac{\sqrt{\pi}}{2^{(1/n)-1}}\Gamma(1/n)} \quad \left[\because \Gamma(m)\Gamma\left(m+\frac{1}{2}\right) = \frac{\sqrt{\pi}}{2^{2m-1}}\Gamma(2m), \text{by} 11\right]$$

$$= \frac{a}{n} \cdot 2^{(1/n)-1} \frac{[\Gamma(1/2n)]^2}{\Gamma(1/n)}.$$

Example 14:

Show that

(i) $\int_0^{\infty} xe^{-ax} \cos \beta x\, dx = \frac{(\alpha^2 - \beta^2)}{(\alpha^2 + \beta^2)^2}$

and (ii) $\int_0^{\infty} xe^{-\alpha x} \sin \beta x\, dx = \frac{2\alpha\beta}{(\alpha^2 + \beta^2)^2}$.

Solution:

We have $\int_0^{\infty} xe^{-\alpha x} \cos \beta x\, dx + i\int_0^{\infty} xe^{-\alpha x} \sin \beta x\, dx$

$$= \int_0^{\infty} xe^{-\alpha x}(\cos \beta x + i \sin \beta x)dx = \int_0^{\infty} xe^{-\alpha x}e^{i\beta x}dx$$

$$= \int_0^{\infty} xe^{-(\alpha - i\beta)x}dx = \int_0^{\infty} e^{-(\alpha - i\beta)x}x^{2-1}dx$$

$$= \frac{\Gamma(2)}{(\alpha - i\beta)^2} = \frac{1}{(\alpha - i\beta)^2} = \frac{(\alpha + i\beta)^2}{[(\alpha - i\beta)(\alpha + i\beta)]^2}$$

$$= \frac{(\alpha^2 - \beta^2) + 2i\alpha\beta}{(\alpha^2 + \beta^2)^2}$$

$$= \frac{\alpha^2 - \beta^2}{(\alpha^2 + \beta^2)^2} + i\frac{2\alpha\beta}{(\alpha^2 + \beta^2)^2}. \quad ...(1)$$

Equating real and imaginary parts in (1), we get

$$\int_0^\infty xe^{-\alpha x} \cos\beta x \, dx = \frac{\alpha^2 - \beta^2}{(\alpha^2 + \beta^2)^2}$$

and $\int_0^\infty xe^{-\alpha x} \sin\beta x \, dx = \frac{2\alpha\beta}{(\alpha^2 + \beta^2)^2}.$

Example 15:

Evaluate

$$\int_0^\infty x^{m-1} \cos bx \, dx \textit{ and } \int_0^\infty x^{m-1} \sin bx \, dx.$$

Solution:

We have

$$\int_0^\infty x^{m-1} \cos bx \, dx - i\int_0^\infty x^{m-1} \sin bx \, dx$$

$$= \int_0^\infty x^{m-1}(\cos bx - i\sin bx)dx = \int_0^\infty x^{m-1}e^{-ibx}dx \int_0^\infty$$

$$= \frac{\Gamma(m)}{(ib)^m} = \frac{\Gamma(m)}{b^m i^m} = \frac{\Gamma(m)}{b^m\left(\cos\frac{1}{2}\pi + i\sin\frac{1}{2}\pi\right)^m}$$

$$= \frac{\Gamma(m)}{b^m}\left(\cos\frac{1}{2}\pi + i\sin\frac{1}{2}\pi\right)^{-m} = \frac{\Gamma(m)}{b^m}\left(\cos\frac{m\pi}{2} - i\sin\frac{m\pi}{2}\right). \quad ...(1)$$

Equating real and imaginary parts in (1), we get

$$\int_0^\infty x^{m-1} \cos bx \, dx = \frac{\Gamma(m)}{b^m}\cos\frac{m\pi}{2}$$

and $\int_0^\infty x^{m-1} \sin bx \, dx = \frac{\Gamma(m)}{b^m}\sin\frac{m\pi}{2}.$

Example 16:

Prove that $\int_{-\infty}^\infty \cos\left(\frac{1}{2}\pi x^2\right)dx = 1.$

Solution:

We have

$$I = \int_{-\infty}^\infty \cos\left(\frac{1}{2}\pi x^2\right)dx = 2\int_0^\infty \cos\left(\frac{1}{2}\pi x^2\right)dx.$$

Put $\frac{1}{2}\pi x^2 = t$ *i.e.*, $x = \sqrt{(2/\pi)}\, t^{1/2}$.

so that $dx = \frac{1}{2}\sqrt{(2/\pi)}\, t^{-1/2} dt$.

$$\therefore I = 2.\frac{1}{2}.\sqrt{\left(\frac{2}{\pi}\right)}\int_0^\infty t^{-1/2}\cos t\, dt = \sqrt{\left(\frac{2}{\pi}\right)}\int_0^\infty t^{-1/2}\cos t\, dt$$

$$= \text{real part in } \sqrt{\left(\frac{2}{\pi}\right)}\int_0^\infty t^{-1/2}e^{-it}dt, \; [\because e^{-it} = \cos t - i\sin t]$$

$$= \text{real part in } \sqrt{\left(\frac{2}{\pi}\right)}\int_0^\infty t^{(1/2)-1}e^{-it}dt$$

$$= \text{real part in } \sqrt{\left(\frac{2}{\pi}\right)}\frac{\Gamma\left(\frac{1}{2}\right)}{(i)^{1/2}}$$

$$= \text{real part in } \sqrt{\left(\frac{2}{\pi}\right)}\frac{\sqrt{\pi}}{\left(\cos\frac{1}{2}\pi + i\sin\frac{1}{2}\pi\right)^{1/2}}$$

$$= \text{real part in } \sqrt{2}.\left(\cos\frac{1}{2}\pi + i\sin\frac{1}{2}\pi\right)^{-1/2}$$

$$= \text{real part in } \sqrt{2}.\left(\cos\frac{1}{4}\pi - i\sin\frac{1}{4}\pi\right)^{-1/2}$$

$$= \sqrt{2}.\cos\frac{\pi}{4} = \sqrt{2}.\frac{1}{\sqrt{2}} = 1.$$

Example 17:

Prove that

$$\int_0^{\pi/2}\frac{\sin^{2m-1}\theta\cos^{2n-1}\theta\, d\theta}{(a\sin^2\theta + b\cos^2\theta)^{m+n}} = \frac{\Gamma(m)\Gamma(n)}{2a^m b^m\Gamma(m+n)}.$$

Solution:

We have $I = \int_0^{\pi/2}\frac{\sin^{2m-1}\theta\cos^{2n-1}\theta\, d\theta}{(a\sin^2\theta + b\cos^2\theta)^{m+n}}$,

$$= \int_0^{\pi/2}\frac{\tan^{2m-1}\theta\sec^2\theta\, d\theta}{(b + a\tan^2\theta)^{m+n}},$$

dividing the Nr. and Dr. by $(\cos^2\theta)^{m+n}$ *i.e.*, by $(\cos\theta)^{2m+2n}$

$$= \int_0^{\pi/2} \frac{\tan^{2m-2}\theta \tan\theta \sec^2\theta \, d\theta}{(b + a\tan^2\theta)^{m+n}}$$

$$= \int_0^{\pi/2} \frac{(\tan^2\theta)^{m-1} \tan\theta \sec^2\theta \, d\theta}{(b + a\tan^2\theta)m + n}$$

Put $a \tan^2\theta = by$ so that $2a \tan\theta \sec^2\theta \, d\theta = b\, dy$.

$$\therefore I = \int_0^{\infty} \frac{(by/a)^{m-1}}{(b + by)^{m+n}} \cdot \frac{b}{2a} dy = \frac{b^m}{b^{m+n}.a^m} \cdot \frac{1}{2} \int_0^{\infty} \frac{y^{m-1}}{(1+y)^{m+n}} dy$$

$$= \frac{1}{2a^m b^n} B(m, n) = \frac{1}{2a^m b^{n2}} \frac{\Gamma(m)\Gamma(n)}{\Gamma(m+n)}.$$

Example 18:

Prove that

$$\int_0^1 \frac{x^{m-1}(1-x)^{n-1}}{(a+x)^{m+n}} dx = \frac{B(m,n)}{a^n (1+a)^m}.$$

Solution:

We know that $B(m, n) = \int_0^1 y^{m-1}(1-y)^{n-1} dy$. ...(1)

Put $y = \dfrac{(1+a)x}{a+x} = (1 + a)\left(1 - \dfrac{a}{a+x}\right)$ so that $dy = \dfrac{a(1+a)}{(a+x)^2} dx$.

When $y = 1$, $x = 1$ and when $y = 0$, $x = 0$.

Also $1 - y = 1 - \dfrac{x(1+a)}{a+x} = \dfrac{a(1-x)}{a+x}$.

$$\therefore B(m, n) = \int_0^1 \frac{(1+a)^{m-1} x^{m-1}}{(a+x)^{m-1}} \cdot \frac{a^{n-1}(1-x)^{n-1}}{(a+x)^{n-1}} \cdot \frac{a(1+a)}{(a+x)^2} dx$$

$$= (1 + a)^m a^n \int_0^1 \frac{x^{m-1}(1-x)^{n-1}}{(a+x)^{m+n}} dx.$$

$$\int_0^1 \frac{x^{m-1}(1-x)^{n-1}}{(a+x)^{m+n}} dx = \frac{B(m,n)}{a^n(1+a)^m} = \frac{\Gamma(m)\Gamma(n)}{a^n(1+a)^m \Gamma(m+n)}.$$

Example 19:

Prove that

$$\int_0^{\infty} e^{-x^2} x^{2n} dx = \sqrt{\pi} \cdot \frac{1.3.5\ldots(2n-1)}{2.4.6\ldots 2n} \int_0^{\infty} e^{x^2} x^{2n+1} dx.$$

Solution:

$$\text{Let } U = \frac{\int_0^\infty e^{-x^2} x^{2n} dx}{\int_0^\infty e^{-x^2} x^{2n+1} dx} = \frac{\int_0^\infty e^{-x^2} . x^{2n-1} x\, dx}{\int_0^\infty e^{-x^2} . x^{2n} x\, dx}.$$

Put $x^2 = t$ so that $2x\, dx = dt$ and $x = t^{1/2}$.

$$\therefore U = \frac{\int_0^\infty e^{-t} (t^{1/2})^{2n-1} dt}{\int_0^\infty e^{-t} (t^{1/2})^{2n} dt} = \frac{\int_0^\infty e^{-t} t^{n-(1/2)} dt}{\int_0^\infty e^{-t} t^{n} dt}$$

$$= \frac{\int_0^\infty e^{-t} t^{\{n+(1/2)\}-1} dt}{\int_0^\infty e^{-t} t^{(n+1)-1} dt} = \frac{\Gamma\left(n+\frac{1}{2}\right)}{\Gamma(n+1)} = \frac{\Gamma\left(\frac{2n+1}{2}\right)}{\Gamma(n+1)}$$

$$= \frac{\left(\frac{2n-1}{2}\right)\left(\frac{2n-3}{2}\right)\left(\frac{2n-5}{2}\right)\cdots\frac{3}{2}\cdot\frac{1}{2}\cdot\Gamma\left(\frac{1}{2}\right)}{n(n-1)(n-2)\ldots 2.1\Gamma(1)}$$

$$= \sqrt{\pi}\cdot\frac{1.3.5\ldots(2n-1)}{2.4.6\ldots(2n)}.$$

$$\therefore \int_0^\infty e^{-x^2} x^{2n} dx = \sqrt{\pi}\cdot\frac{1.3.5\ldots(2n-1)}{2.4.6\ldots(2n)} \int_0^\infty e^{x^2} x^{2n+1} dx.$$

Example 20:

Prove that

$$\int_0^p x^m (p^{q-xq})^n dx = \frac{p^{qn+m+1}}{q} B\left(n+1, \frac{m+1}{q}\right),$$

if $p > 0$, $q > 0$, $m + 1 > 0$, $n + 1 > 0$.

Solution:

Let $I = \int_0^p x^m (p^{q-xq})^n dx$.

Put $x^q = p^q y$ or $x\ py^{1/q}$,

so that $dx = p\ .(1/q)\ y^{(1/q)-1}\ dy$.

When $x = 0$, $y = 0$

and when $x = p$, $y = 1$.

$$\therefore I = \int_0^1 (py^{1/q})^m (p^q - p^q y)^n \cdot \frac{p}{q} y^{(1/q)-1} dy$$

$$= \text{pm.pqn} \cdot \frac{p}{q}\int_0^1 y^{(m/q)+(1/q)-1}(1-y)^n\,dy$$

$$= \frac{p^{qn+m+1}}{q}\int_0^1 y^{[(m+1)/q]-1}(1-y)^{(n+1)-1}\,dy$$

$$= \frac{p^{qn+m+1}}{q}B\left(\frac{m+1}{q}, n+1\right) = \frac{p^{qn+m+1}}{q}B\left(n+1, \frac{m+1}{q}\right),$$

if $m+1>0$ and $n+1>0$.

Example 21:

Prove that

$$\int_0^\pi \frac{\sin^{n-1} x\,dx}{(a+b\cos x)^n} = \frac{2^{n-1}}{(a^2-b^2)^{n/2}}\cdot B\left(\frac{n}{2},\frac{n}{2}\right),\ a>b.$$

Solution:

$$\text{Let } I = \int_0^\pi \frac{\sin^{n-1} x}{(a+b\cos x)^n}dx$$

$$= \int_0^\pi \frac{\left(2\sin\frac{1}{2}x\cos\frac{1}{2}x\right)^{n-1}}{\left[a\left(\cos^2\frac{1}{2}x+\sin^2\frac{1}{2}x\right)+b\left(\cos^2\frac{1}{2}x-\sin^2\frac{1}{2}x\right)\right]^n}dx$$

$$= \int_0^\pi \frac{2^{n-1}\left(\sin\frac{1}{2}x\right)^{n-1}\left(\cos\frac{1}{2}x\right)^{n-1}}{\left[(a+b)\cos^2\frac{1}{2}x+(a-b)\sin^2\frac{1}{2}x\right]^n}dx$$

$$= \int_0^\pi \frac{2^{n-1}\left(\tan\frac{1}{2}x\right)^{n-1}\sec^2\frac{1}{2}x\,dx}{\left[(a+b)+(a-b)\tan^2\frac{1}{2}x\right]^n},$$

dividing the Nr. and Dr. by $\left(\cos^2\frac{1}{2}x\right)^n$ *i.e.,* $\left(\cos^2\frac{1}{2}x\right)^{2n}$

$$= 2^{n-1}\int_0^\pi \frac{\left(\tan\frac{1}{2}x\right)^{n-2}\tan\frac{1}{2}x\sec^2\frac{1}{2}x\,dx}{\left[(a+b)+(a-b)\tan^2\frac{1}{2}x\right]^n}$$

Put $(a-b)\tan^2\frac{1}{2}x = (a+b)\,y$ so that

$$(a-b).\left(2\tan\frac{1}{2}x\sec^2\frac{1}{2}x\right)\cdot\frac{1}{2}dx = (a+b)\,dy.$$

When $x = 0$, $y = 0$

and when $x \to \pi$, $y = \to \infty$.

$$\therefore I = 2^{n-1}\int_0^\infty \frac{\{(a+b)y/(a-b)\}^{(n-2)/2}}{[(a+b)+(a+b)y]^n}\frac{a+b}{a-b}dy$$

$$= 2^{n-1}\int_0^\infty \frac{(a+b)^{(n-2)/2}(a+b)}{(a+b)^n(a-b)^{(n-2)/2}(a-b)}\cdot\frac{y^{(n-2)/2}}{(a+y)^n}dy$$

$$= \frac{2^{n-1}}{(a+b)^{n/2}(a-b)^{n/2}}\int_0^\infty \frac{y^{(n/2)-1}}{(1+y)^{(n/2)+(n/2)}}dy$$

$$= \frac{2^{n-1}}{(a^2-b^2)^{n/2}}B\left(\frac{n}{2},\frac{n}{2}\right).$$

Example 22:

Prove that

$$\int_0^\pi \frac{\sqrt{(\sin x)}}{(5+3\cos x)^{3/2}}dx = \frac{\left[\Gamma\left(\frac{3}{4}\right)\right]^2}{2\sqrt{(2\pi)}}$$

Solution:

Do your self.

Example 23:

Show that $\displaystyle\int_0^\infty \frac{x^c}{c^x}dx = \frac{\Gamma(c+1)}{(\log c)^{c+1}}$, *c > 1.*

Solution:

$$\text{We have } I = \int_0^\infty \frac{x^c}{c^x}dx = \int_0^\infty \frac{x^c}{e^{\log c^x}}dx$$

$$= \int_0^\infty \frac{x^c}{e^{x\log c}}dx = \int_0^\infty e^{-x\log c}x^c dx.$$

Put $x\log c = y$ so that $(\log c)\,dx = dy$.

When $x = 0$, we have $y = 0$

and when $x \to \infty$, $y \to \infty$

because $c > 1 \Rightarrow \log c > 0$.

$$\therefore I = \int_0^\infty e^{-y}\left(\frac{y}{\log c}\right)^c \frac{dy}{\log c}$$

$$= \frac{1}{(\log c)^{c+1}}\int_0^\infty e^{-y} y^{(c+1)-1} dy$$

$$= \frac{1}{(\log c)^{c+1}}\Gamma(c+1),$$

provided $c + 1 > 0$ which is so because $c > 1$.

Example 24:

Prove that $\int_0^1 x^{n-1}\left(\log\frac{1}{x}\right)^{m-1} dx = \frac{\Gamma m}{n^m}$, *(m, n > 0).*

Solution:

Let $I = \int_0^1 x^{n-1}\left(\log\frac{1}{x}\right)^{m-1} dx$.

Put $\log(1/x) = y$

i.e., $1/x = e^y$ *i.e.*, $x = e^{-y}$

so that $dx = -e^{-y} dy$.

Also when $x \to 0$, $y \to \infty$

and when $x = 1$, $y = 0$. [Note tht $\log \infty = \infty$]

$$\therefore I = -\int_\infty^0 (e^{-y})^{n-1} y^{m-1} e^{-y} dy$$

$$= \int_0^\infty (e^{-y})^{n-1+1} y^{m-1} dy = \int_0^\infty e^{-ny} y^{m-1} dy$$

$$= \frac{\Gamma m}{n^m}, \text{ provided } m > 0 \text{ and } n > 0.$$

Example 25:

Prove that $B(n, n) = \dfrac{\sqrt{\pi}\,\Gamma(n)}{2^{2n-1}\Gamma\left(n+\frac{1}{2}\right)}$.

Solution:

By the definition of Beta function, we have

$$B(n, n) = \int_0^1 x^{n-1}(1-x)^{n-1} dx, \; n > 0.$$

Now put $x = \sin^2\theta$ and proceed as in 11 to get the required result.

Example 26:

Evaluate the following integral:

(i) $\int_0^{\pi/2} \sqrt{(\tan\theta)}d\theta$

(ii) $\int_0^{\pi/2} \sqrt{(\cot\theta)}d\theta$

(iii) $\int_0^{\infty} \frac{dy}{1+y^4}$.

Solution:

(i) We have $I = \int_0^{\pi/2} \sqrt{(\tan\theta)}d\theta$

$$= \int_0^{\pi/2} \frac{\sin^{1/2}\theta}{\cos^{1/2}\theta}d\theta = \int_0^{\pi/2} \sin^{1/2}\theta\cos^{-1/2}\theta\, d\theta$$

$$= \frac{\Gamma\left\{\frac{1}{2}\left(\frac{1}{2}+1\right)\right\}\Gamma\left\{\frac{1}{2}\left(-\frac{1}{2}+1\right)\right\}}{2\Gamma\left\{\frac{1}{2}\left(\frac{1}{2}-\frac{1}{2}+2\right)\right\}} = \frac{\Gamma\left(\frac{3}{4}\right)\Gamma\left(\frac{1}{4}\right)}{2\,\Gamma 1}$$

$$= \frac{\Gamma\left(\frac{1}{4}\right)\Gamma\left(1-\frac{1}{4}\right)}{2} = \frac{1}{2}\cdot\frac{\pi}{\sin\frac{1}{4}\pi} = \frac{\pi}{2.(1/\sqrt{2})} = \frac{\pi}{\sqrt{2}}.$$

(ii) We have $\int_0^{\pi/2} \sqrt{(\cot\theta)}d\theta$

$$= \int_0^{\pi/2} \cos^{1/2}\theta\sin^{-1/2}\theta\, d\theta = \frac{\pi}{\sqrt{2}},$$

proceeding as in part (i) of this exercise.

(iii) Let $I = \int_0^{\infty} \frac{dy}{1+y^4}$.

Put $y = \tan\theta$

i.e., $y = \sqrt{(\tan\theta)}$

so that $dy = \frac{1}{2}\tan^{-1/2}\theta\sec^2\theta\, d\theta$.

When $y = 0, \theta = 0$

and when $y \to \infty, \theta \to \pi/2$.

$$\therefore I = \int_0^{\pi/2} \frac{1}{1+\tan^2\theta} \cdot \frac{1}{2} \tan^{-1/2}\theta \sec^2\theta \, d\theta$$

$$= \frac{1}{2}\int_0^{\pi/2} \tan^{-1/2}\theta \, d\theta = \frac{1}{2}\int_0^{\pi/2} \sin^{-1/2}\theta \cos^{1/2}\theta \, d\theta$$

$$= \frac{1}{2} \cdot \frac{\pi}{2} = \frac{\pi}{2\sqrt{2}},$$ proceeding as in part (i) of this exercise.

Example 27:

Evaluate the following integrals:

(i) $\int_0^{\infty} \frac{y^{q-1}}{(1+y)^{p+q}} dy$

(ii) $\int_0^{\infty} 4x^4 e^{-x^4} dx$

(iii) $\int_{-\infty}^{\infty} e^{-k^2 d^2} dx$

Solution:

(i) Let $I = \int_0^{\infty} \frac{y^{q-1}}{(1+y)^{p+q}} dy$.

Put $y = \frac{1}{x} - 1 = \frac{1-x}{x}$

so that $dy = -\frac{1}{x^2} dx$.

Also $\frac{1}{x} = y + 1$ or $x = \frac{1}{y+1}$.

When $y = 0$, we have $x = 1$

and when $y \to \infty$, $x \to 0$.

$$\therefore I = \int_1^0 \left(\frac{1-x}{x}\right)^{q-1} . x^{p+q} . \left(-\frac{1}{x^2}\right) dx$$

$$= \int_0^1 x^{p+q} (1-x)^{q-1} \cdot \frac{1}{x^{q-1}} \cdot \frac{1}{x^2} dx$$

$$= \int_0^1 x^{p-1} (1-x)^{q-1} dx$$

$= B(p, q)$, (provided $p > 0$ and $q > 0$)

$$= \frac{\Gamma p \Gamma q}{\Gamma(p+q)}.$$

(ii) Let $I = \int_0^\infty 4x^4 e^{-x^4}\,dx$.

Put $x^4 = t$ *i.e.*, $x = t^{1/4}$

so that $dx = \frac{1}{4} t^{-3/4} dt$.

Then $I = \int_0^\infty 4.t.e^{-t}.\frac{1}{4} t^{-3/4} dt$

$= \int_0^\infty e^{-t} t^{1/4} dt = \int_0^\infty e^{-t} t^{(5/4)-1} dt$

$= \Gamma\,(5/4) = \frac{1}{4}\Gamma\left(\frac{1}{4}\right)$.

(iii) Let $I = \int_{-\infty}^\infty e^{-k^2x^2}\,dx$.

Since $e^{-k^2x^2}$ is an even function of x, therefore

$I = 2\int_0^\infty e^{-k^2x^2 dx}$.

Now put $k^2 x^2 = t$ *i.e.*, $x = t^{1/2}$

so that $k\,dx = \frac{1}{2} t^{(1/2)-1} dt$.

When $x = 0$, $t = 0$

and when $x \to \infty$, $t \to \infty$.

$\therefore I = 2\int_0^\infty e^{-t.} \frac{1}{2k} t^{(1/2)-1} dt$

$= \frac{1}{k}\int_0^\infty e^{-t} t^{(1/2)-1} dt = \frac{1}{k}\Gamma\left(\frac{1}{2}\right) = \frac{1}{k}\cdot\sqrt{\pi}$.

Example 28:

Evaluate $\iint x^{2l-1} y^{2m-1} dx\,dy$ *for all positive values of x and y such that* $x^2 + y^2 \le c^2$.

Solution :

Let us denote the given integral by I. Then we have to find the value of I extended to all positive values of x and y subject to the condition

$$\left(\frac{x}{c}\right)^2 + \left(\frac{y}{c}\right)^2 \le 1.$$

Put $(x/c)^2 = u$ *i.e.*, $x = cu^{1/2}$, so that $dx = \frac{1}{2} cu^{-1/2} du$,

and $(y/c)^2 = v$ *i.e.*, $y = cv^{1/2}$, so that $dy = \frac{1}{2} cv^{-1/2} dv$.

Then the required integral

$$I = \iint (cu^{1/2})^{2l-1} (cv^{1/2})^{2m-1} . \frac{1}{2} cu^{-1/2} . \frac{1}{2} cv^{-1/2} du\, dv$$

$$= \frac{1}{4} c^{2l+2m} \iint u^{l-1} v^{m-1} du\, dv, \text{ where u, v take all +ive values}$$

subject to the condition $u + v \leq 1$

$$= \frac{1}{4} c^{2l+2m} . \frac{\Gamma(l)\Gamma(m)}{\Gamma(l+m+1)}, \text{ by Dirichlet's theorem.}$$

Exemple 29 :

Find the value of

$$\iint \ldots \int dx_1 dx_2 \ldots dx_n$$

extended to all positive values of the variables, subject to the condition

$$x_1^2 + x_2^2 + \ldots + x_n^2 < R^2.$$

Solution :

Let us denote the given integral by I. Then we have to find the value of I extended to all positive values of $x_1, x_2, \ldots x_n$ subject to the condition

$$\frac{x_1^2}{R^2} + \frac{x_2^2}{R^2} + \ldots + \frac{x_n^2}{R^2} < 1.$$

Put $(x_1/R)^2 = u_1$ *i.e.*, $x_1 = Ru_1^{1/2}$, so that $dx_1 = \frac{1}{2} Ru_1^{-1/2} du_1$,

$(x_2/R)^2 = u_2$ *i.e.*, $x_2 = Ru_2^{1/2}$, so that $dx_2 = \frac{1}{2} Ru_2^{-1/2} du_2$, and so on.

Then the required integral

$$I = \iint \ldots \int \left(\frac{1}{2}\right)^n R^n u_1^{-1/2} u_2^{-1/2} \ldots u_n^{-1/2} du_1 du_2 \ldots du_n$$

$$= \left(\frac{R}{2}\right)^n \iint \ldots \int u_1^{(1/2)-1} u_2^{(1/2)-1} \ldots u_n^{(1/2)-1} du_1 du_2 \ldots du_n,$$

subject to the condition $u_1 + u_2 + \ldots + u_n < 1$

$$= \left(\frac{R}{2}\right)^n \frac{\left\{\Gamma\left(\frac{1}{2}\right)\right\}^n}{\Gamma\left(1+n.\frac{1}{2}\right)}, \text{ by Dirichlet's theorem}$$

$$= \left(\frac{R}{2}\right)^n \cdot \frac{\pi^{n/2}}{\Gamma\left(1+\frac{1}{2}n\right)}. \qquad \left[\because \Gamma\left(\frac{1}{2}\right) = \sqrt{\pi}\right]$$

Example 30:

Find the volume of the solid surrounded by the surface

$(x/a)^{2/3} + (y/b)^{2/3} + (z/c)^{2/3} = 1.$

Solution :

Since the equation $(x/a)^{2/3} + (y/b)^{2/3} + (z/c)^{2/3} = 1$ does not changeby putting $-x$ for x, $-y$ for y and $-z$ for z, therefore the surface represented by this equation is symmetrical in all the eight octants.

So the volume of the solid surrounded by this surface = 8 × the volume of the portion of this solid lying in the positive octant.

Now the volume of a small element situated at any point

$(x, y, z) = dx\, dy\, dz.$

∴ the volume of the solid in the positive octant

$$= \iiint dx\, dy\, dz,$$

where the integral is extended to all positive values of the variables x, y, z subject to the condition

$(x/a)^{2/3} + (y/b)^{2/3} + (z/c)^{2/3} \le 1.$

Now put $(x/a)^{2/3} = u$, $(y/b)^{2/3} = v$, $(z/c)^{2/3} = w$

i.e., $x = au^{3/2}$, $y = bv^{2/3}$, $z = cw^{3/2}$

so that $dx = \frac{3}{2} au^{1/2} du, dy = \frac{3}{2} dv^{1/2} dv, dz = \frac{3}{2} cw^{1/2} dw.$

= the volume in the positive octant

$$= \iiint \frac{27}{8} abc\, u^{(3/2)-1} v^{(3/2)-1} w^{(3/2)-1} du\, dv\, dw,$$

where $u + v + w \le 1$

$$= \frac{27}{8} abc \frac{[\Gamma(3/2)]^3}{\Gamma\left(\frac{9}{2}+1\right)} = \frac{27}{8} abc \cdot \frac{\left(\frac{1}{2}.\sqrt{\pi}\right)^3}{\frac{9}{2}\cdot\frac{7}{2}\cdot\frac{5}{2}\cdot\frac{3}{2}\cdot\frac{1}{2}\sqrt{\pi}}$$

$$= \frac{27}{8} abc \cdot \frac{\pi}{8} \cdot \frac{32}{27.35} = \frac{\pi\, abc}{8} \cdot \frac{4}{35}.$$

Hence the required volume $= 8 \cdot \frac{\pi abc}{8} \cdot \frac{4}{35} = \frac{4\pi abc}{35}.$

Example 31:

Find the volume enclosed by the surface

$(x/a)^{2n} + (y/b)^{2n} + (z/c)^{2n} = 1.$

Solution:

The given surface is symmetrical in all the eight octants.

The volume V in the positive octant

$= \iiint dx\, dy\, dz,$ where the integral is extended to all positive values of the variables x, y, z subject to the condition

$(x/a)^{2n} + (y/b)^{2n} + (z/c)^{2n} \le 1.$

Now put $(x/a)^{2n} = u$, $(y/b)^{2n} = v$, $(z/c)^{2n} = w$

i.e., $x = au^{1/2n}$, $y = bv^{1/2n}$, $z = cw^{1/2n}$

so that $dx = \frac{a}{2n} u^{(1/2n)-1} du$, etc.

$\therefore V = \frac{abc}{8n^3} \iiint u^{(1/2n)-1} v^{(1/2n)-1}\ w^{(1/2n)}\ du\, dv\, dw$, where $u + v + w \le 1$

$$= \frac{abc}{8n^3} \cdot \frac{[\Gamma(1/2n)]^3}{\Gamma\{(3/2n)+1\}} = \frac{abc}{8n^3} \cdot \frac{[\Gamma(1/2n)]^3}{(3/2n).\Gamma(3/2n)}$$

$$= \frac{abc}{8n^3} \cdot \frac{[\Gamma(1/2n)]^3}{(3/2n).\Gamma(3/2n)} \frac{abc}{12n^2} \cdot \frac{[\Gamma(1/2n)]^3}{\Gamma(3/2n)}.$$

Hence the total volume encolsed by the given surface

$$= 8V = \frac{2}{3} \cdot \frac{abc}{n^2} \cdot \frac{[\Gamma(1/2n)]^3}{\Gamma(3/2n)}.$$

Example 32:

Show that the integral $\iiint x^{l-1} y^{m-1} z^{n-1} dx\, dy\, dz$ *integrated over the region in the first octant below the surface* $(x/a)^p + (y/b)^q + (z/c)^r = 1$ *is*

$$\frac{a^l b^m c^n}{pqr} \cdot \frac{\Gamma(l/p)\Gamma(m/q)\Gamma(n/r)}{\Gamma(l/p+m/q+n/r+1)}.$$

Or

Evaluate the integral $\iiint x^{l-1} y^{m-1} z^{n-1} dx\,dy\,dz$, *where x, y, z are all positive but limited by the condition*

$(x/a)^p + (y/b)^q + (z/c)^r \le 1.$

Solution:

The required integral $= \iiint x^{l-1} y^{m-1} z^{n-1} dx\,dy\,dz$, where the integral is extended to all positive values of the variables x, y and z subject to the condition

$(x/a)^p + (y/b)^q + (z/c)^r \le 1.$

Put $(x/a)^p = u$ *i.e.*, $x = au^{1/p}$ os that $dx = (a/p)\, u^{(1/p)-1} du$,

$(y/b)^q = v$ *i.e.*, $y = bv^{1/q}$ so that $dy = (b/q)\, v^{(1/q)-1} dv$,

and $(z/c)^r = w$ *i.e.*, $z = cw^{1/r}$ so that $dz = (c/r)\, w^{(1/r)-1} dw$.

Then the required integral

$$= \iiint (a^{l-1} u^{(l-1)/p})(b^{m-1} v^{(m-1)/q})(c^{n-1} w^{(n-1)/r}).$$

$$\frac{a}{p} u^{(1/p)-1} \cdot \frac{b}{q} v^{(1/q)-1} \cdot \frac{c}{r} w^{(1/r)-1} du\,dv\,dw$$

$$= \frac{a^l b^m c^n}{pqr} \iiint u^{(l/p)-1} v^{(m/q)-1} w^{(n/r)-1} du\,dv\,dw, \text{ where } u + v + w \le 1$$

$$= \frac{a^l b^m c^n}{pqr} \cdot \frac{\Gamma(l/p)\Gamma(m/q)\Gamma(n/r)}{\Gamma(l/p+m/q+n/r+1)}, \text{ by Dirichlet's integral.}$$

Example 33:

Show that if l, m, n are all positive

$$\iiint x^{l-1} y^{m-1} z^{n-1} dx\,dy\,dz = \frac{a^l b^m c^n}{8} \cdot \frac{\Gamma(l/2)\Gamma(m/2)\Gamma(n/2)}{\Gamma(l/2+m/2+n/2+1)}$$

where the triple integral is taken throughout the part of the ellipsoid $(x^2/a^2) + (y^2/b^2) + (z^2/b^2) = 1$, which lies in the positive octant.

Solution:

The required integral

$$= \iiint x^{l-1} y^{m-1} z^{n-1} dx\, dy\, dz,$$

where the integral is extended to all positive values of the variables x, y, z subject to the condition

$$x^2/a^2 + y^2/b^2 + z^2/c^2 \leq 1.$$

Put $x^2/a^2 = u$ *i.e.*, x $au^{1/2}$ so that $dx = \frac{1}{2} a u^{-1/2} du$,

$y^2/b^2 = v$ *i.e.*, $y = bv^{1/2}$ so that $dy = \frac{1}{2} b v^{-1/2} dv$,

and $z^2/c^2 = w$ *i.e.*, $z = cw^{1/2}$ so that $dz = \frac{1}{2} c w^{-1/2} dw$.

Then the required integral

$$= \iiint a^{l-1} u^{(l-1)} . b^{m-1} v^{(m-1)/2} . c^{n-1} w^{(n-1)/2}$$

$$\times \frac{1}{8} abc u^{-1/2} v^{-1/2} w^{-1/2} du\, dv\, dw$$

$$= \frac{a^l b^m c^n}{8} \iiint u^{(l/2)-1} v^{(m/2)-1} w^{(n/2)-1} du\, dv\, dw, \quad \text{where } u + v + w \leq 1$$

$$= \frac{a^l b^m c^n}{8} \frac{\Gamma(l/2)\Gamma(m/2)\Gamma(n/2)}{\Gamma(l/2 + m/2 + n/2 + 1)}. \quad \text{(by Dirichlet's integral)}$$

Example 34:

Evaluate $\iiint dx\, dy\, dz$, where $\frac{x^2}{a^2} + \frac{y^2}{b^2} + \frac{z^2}{c^2} \leq 1.$

Solution:

Here we are to evaluate the given integral over the whole volume of the ellipsoid $x^2/a^2 + y^2/b^2 + z^2/c^2 = 1$ which is symmetrical in all the eight octants. Obviously the required integral

$$= 8 \iiint dx\, dy\, dz$$

where the integral is to be evaluated throughout the volume of the ellipsoid which lies in the positive octant.

Let $I = \iiint dx\, dy\, dz$, where the integral is extended to all positive values of the variables x, y, z subject to the condition

$$x^2/a^2 + y^2/b^2 + z^2/c^2 \leq 1.$$

Put $x^2/a^2 = u$ *i.e.*, $x = au^{1/2}$

so that $dx = \frac{1}{2} a u^{-1/2} du$,

$y^2/b^2 = v$ *i.e.*, $y = bv^{1/2}$ so that $dy = \frac{1}{2} bv^{-1/2} dv$,

and $z^2/c^2 = w$ *i.e.*, $z = cw^{1/2}$ so that $dz = \frac{1}{2} cw^{-1/2} dw$.

Then $I = \iiint \frac{1}{8} abc\, u^{-1/2} v^{-1/2} w^{-1/2} du\, dv\, dw$

$$= \iiint \frac{1}{8} abc\, u^{(1/2)-1} v^{(1/2)-1} w^{(1/2)-1} du\, dv\, dw, \quad \text{where } u + v + w \leq 1$$

$$= \frac{abc}{8} \frac{\Gamma\left(\frac{1}{2}\right).\Gamma\left(\frac{1}{2}\right).\Gamma\left(\frac{1}{2}\right)}{\Gamma\left(\frac{1}{2}+\frac{1}{2}+\frac{1}{2}+1\right)}, \qquad \text{(by Dirichlet's integral)}$$

$$= \frac{abc}{8} \frac{(\sqrt{\pi})^3}{\Gamma\left(\frac{5}{2}\right)} = \frac{abc}{8} \frac{\pi.\sqrt{\pi}}{\frac{3}{2}.\frac{1}{2}.\sqrt{\pi}} = \frac{\pi abc}{6}.$$

Hence the required integral

$$= 8 \cdot \frac{\pi abc}{6} = \frac{4}{3} \pi\, abc.$$

Example 35:

Find the volume of the ellipsoid $(x^2/a^2) + (y^2/b^2) + (z^2/c^2) = 1$.

Solution:

The given ellipsoid is symmetrical in all the eight octants.

Therefore, the volume of the given ellipsoid = 8 × the volume of the part of the ellipsoid lying in the positive octant.

Now the volume of a small element situated at any point (x, y, z) = dxdydz.

∴ the volume of the given ellipsoid

$$= 8 \iiint dx\, dy\, dz,$$

where the integral is extended to all positive values of the variables x, y, z subject to the condition

$$x^2/a^2 + y^2/b^2 + z^2/c^2 \leq 1.$$

Hence the required volume $= 8 \cdot \frac{\pi abc}{6} = \frac{4}{3} \pi\, abc$.

Example 36:

Find the volume in the positive octant of the ellipsoid

$$x^2/a^2 + y^2/b^2 + z^2/c^2 = 1.$$

Solution:

Do your self. **Ans.** π abc/6.

Example 37:

Find the volume of the sphere $x^2+y^2 + z^2 = a^2$.

Solution:

The given shpere $x^2/y^2 + z^2 = a^2$ is symmetrical in all the eight octant.

$\therefore$ the volume of the given sphere $x^2 + y^2 + z^2 = a^2$

$$= 8\int\int\int dx\, dy\, dz,$$

where the integral is extended to all positive values of the variables x, y, z subject to the condition $\frac{x^2}{a^2}+\frac{y^2}{a^2}+\frac{z^2}{a^2} \leq 1.$

Now put $x = au^{1/2}$, $y = av^{1/2}$, $z = aw^{1/2}$.

The required volume $= \frac{4}{3}\pi a^3$.

Example 38:

Evaluate $\int\int\int xyz\, dx\, dy\, dz$ *for all positive values of the variables throughout the ellipsoid* $x^2/a^2 + y^2/b^2 + z^2/c^2 = 1$.

Solution:

We are to evaluate the integral $\int\int\int xyz\, dx\, dy\, dz$, for all positive values of the variables x, y, z subject to the condition

$$x^2/a^2 + y^2/b^2 + z^2/c^2 \leq 1.$$

Put $x^2/a^2 = u$, $y^2/b^2 = v$, $z^2/c^2 = w$ so that

$$x\, dx = \frac{1}{2}a^2\, du,$$

$$y\, dy = \frac{1}{2}b^2\, dv,$$

$$z\, dz = \frac{1}{2}c^2\, dw.$$

Then the required integral

$$= \frac{1}{8}a^2b^2c^2 \iiint du\,dv\,dw, \text{ where } u + v + w \leq 1$$

$$\frac{1}{8}a^2b^2c^2 \iiint u^{1-1}\, v^{1-1}\, w^{1-1}\, du\,dv\,dw \text{ where } u + v + w \leq 1$$

$$= \frac{1}{8}a^2b^2c^2 \frac{\Gamma(1)\Gamma(1)\Gamma(1)}{\Gamma(1+1+1+1)}, \text{ by Dirichlet's integral}$$

$$= \frac{a^2b^2c^2}{8}\frac{1}{\Gamma(4)} = \frac{a^2b^2c^2}{8} \times \frac{1}{3.2.1} = \frac{a^2b^2c^2}{48}.$$

Example 39:

Find the volume of the tetrahedron bounded by the plane $x/a + y/b + z/c = 1$ and the coordinate planes.

Solution:

The volume of a small element situated at any point (x, y, z) = dx dy dz.

∴ the volume of the given tetrahedron

$= \iiint dx\,dy\,dz$, where the integral is extended throughout the volume enclosed by the coordinate planes and the place $x/a + y/b + z/c = 1$

$= \iiint dx\,dy\,dz$, where the integral is extended to all positive values of the variables x, y, z subject to the condition $x/a + y/b + z/c \leq 1$.

Put $x/a = u$, $y/b = v$, $z/c = w$ so that

$dx = a\,du$, $dy = b\,dv$, $dz = c dw$.

Then the required volume $= \iiint abc\,du\,dv\,dw$, (where $u + v + w \lesssim 1$)

$$= abc \iiint u^{1-1}v^{1-1}w^{1-1}du\,dv\,dw$$

$$= abc \frac{[\Gamma(1)]^3}{\Gamma(1+1+1+1)}, \text{ by Dirichlet's integral}$$

$$= abc \cdot \frac{1}{\Gamma(4)} = \frac{abc}{3.2.1} = \frac{abc}{6}.$$

Example 40:

The plane $x/a + y/b + z/c = 1$ meets the coordinate axes in the points A, B, C Use Dirichlet's integral to evaluate the mass of the tetrahedron OABC, the density at any point (x, y, z) being kxyz.

Solution:

The mass of a small element of volume dx dy dz situated at any point (x, y, z)

$= \rho.\ dx\ dy\ dz,$ where ρ is the density per unit volume

$= k\ xyz\ dx\ dy\ dz.$

$\therefore$ the mass of the tetrahedron OABC

$= \iiint k\ xyz\ dx\ dy\ dz,$ where the integral is extended to all positive values of the variables x, y, z subject to the condition

$(x/a) + (y/b) + (z/c) \le 1.$

Put $x/a = u$, $y/b = v$, $z/c = w$ so that

$dx = a du$, $dy = b dv$, $dz = c dw$.

Then the required mass

$$= \iiint k.au.bv.cw.abc\ du\ dv\ dw, \text{ where } u + v + w \le 1$$

$$= k a^2 b^2 c^2 \iiint u^{2-1} v^{2-1} w^{2-1} du\ dv\ dw$$

$$= k a^2 b^2 c^2 \frac{[\Gamma(2)]^3}{\Gamma(2+2+2+1)}, \text{ by Dirichlet's integral}$$

$$= k a^2 b^2 c^2 \cdot \frac{1}{\Gamma(7)}, \qquad [\because \Gamma(2) = 1\ \Gamma(1) = 1]$$

$$= \frac{k a^2 b^2 c^2}{6.5.4.3.2.1} = \frac{k a^2 b^2 c^2}{720}.$$

Example 41:

Find the value of $\iiint \log(x+y+z) dx\ dy\ dz$, *the integral extending over all positive values of x, y, z subject to the condition* $x + y + z < 1$.

Solution:

Here the integral is to be extended for all positive values of x, y and z such that $0 < x + y + z < 1$.

$\therefore$ the required integral

$$= \iiint \log(x+y+z) dx\ dy\ dz, \text{ where } 0 < x + y + z < 1$$

$$= \iiint \log(x+y+z) x^{1-1} y^{1-1} z^{1-1} dx\ dy\ dz$$

$$= \frac{\Gamma(1)\Gamma(1)\Gamma(1)}{\Gamma(1+1+1+1)}\int_0^1 (\log u)u^{1+1+1-1}du,$$

by Liouville's extension of Dirichlet's theorem

$$= \frac{1}{\Gamma(3)}\int_0^1 u^2 \log u \, du, \qquad [\because \Gamma(1) = 1]$$

$$= \frac{1}{2.1}\left[\left(\log u.\frac{u^3}{3}\right)_0^1 - \int_0^1 \frac{1}{u}\cdot\frac{u^3}{3}du\right]_0^1,$$

integrating by parts taking u^2 as the second function

$$= \frac{1}{2}\left[0 - \frac{1}{3}\lim_{u \to 0} u^3 \log u - \frac{1}{3}\int_0^1 u^2 du\right]$$

$$= -\frac{1}{6}\left[\frac{u^3}{3}\right]_0^1, \qquad \left[\because \lim_{u \to 0} u^3 \log u = 0\right]$$

$$= -\frac{1}{18}.$$

Note:

$$\lim_{u \to 0} u^3 \log u = \lim_{u \to 0}\frac{\log u}{1/u^3}$$

$$= \lim_{u \to 0}\frac{1/u}{-3/u^4} = \lim_{u \to 0} -\frac{1}{3}u^3 = 0.$$

Example 42:

Evaluate $\iiint e^{x+y+z}dx\,dy\,dz$ *taken over the positive octant such that* $x + y + z \le 1$.

Solution:

Here the integral is to be extended for all positive values of x, y and z such that $0 \le x + y + x \le 1$.

$\therefore$ the required integral

$$\iiint e^{x+y+z}dx\,dy\,dz = \iiint e^{x+y+z}x^{1-1}y^{1-1}z^{1-1}dx\,dy\,dz,$$

where $0 \le x + y + z \le 1$

$$= \frac{\Gamma(1)\Gamma(1)\Gamma(1)}{\Gamma(1+1+1)}\int_0^1 e^u u^{3-1}du,$$

by Liouville's extension of Dirichlet's integral

$$= \frac{1.1.1}{2.1}\left[\left\{u^2.e^u\right\}_0^1 - \int_0^1 2u.e^u du\right],$$

integrating by parts taking e^u as the second function

$$=\frac{1}{2}\left[e-2\left\{(e^u.u)_0^1-\int_0^1 1.e^u du\right\}\right]$$

$$=\frac{1}{2}\left[e-2e+2\left\{e^u\right\}_0^1\right]=\frac{1}{2}[e-2e+2e-2]=\frac{(e-2)}{2}.$$

Exemple 43:

Evaluate

$\iiint x^{-1/2}y^{-1/2}z^{-1/2}(1-x-y-z)^{1/2}\,dx\,dy\,dz$ extended to all positive values of the variables subject to the condition $x+y+z<1$.

Solution:

The given condition is

$0<x+y+z<1.$

$\therefore$ the required integral

$$\iiint x^{-1/2}y^{-1/2}z^{-1/2}(1-x-y-z)^{1/2}\,dx\,dy\,dz$$

$$=\iiint x^{1/2-1}y^{1/2-1}z^{1/2-1}\{1-(x-y-z)\}^{1/2}\,dx\,dy\,dz,$$

where $0<x+y+z<1$

$$=\frac{\Gamma\left(\frac{1}{2}\right)\Gamma\left(\frac{1}{2}\right)\Gamma\left(\frac{1}{2}\right)}{\Gamma\left(\frac{1}{2}+\frac{1}{2}+\frac{1}{2}\right)}\int_0^1(1-u)^{1/2}u^{1/2+1/2+1/2-1}du,$$

by Liouville's extension of Dirichlet's theorem

$$=\frac{\left[\Gamma\left(\frac{1}{2}\right)\right]^3}{\Gamma(3/2)}\int_0^1(1-u)^{3/2-1}u^{(3/2)-1}du=\frac{(\sqrt{\pi})^3}{\frac{1}{2}.\sqrt{\pi}}\cdot\frac{\Gamma\frac{3}{2}\Gamma\frac{3}{2}}{\Gamma\left(\frac{3}{2}+\frac{3}{2}\right)}.$$

$$=2\pi\cdot\frac{\frac{1}{2}\sqrt{\pi}.\frac{1}{2}\sqrt{\pi}}{2.1}=\frac{\pi^2}{4}.$$

Exemple 44:

Prove that $\iiint\frac{dx\,dy\,dz}{\sqrt{(1-x^2-y^2-z^2)}}=\frac{\pi^2}{8}$, *the integral being extended to all positive values of the variables for which the expression is real.*

Solution:

The given expression $1/\sqrt{(1-x^2-y^2-z^2)}$ is real if

Therefore the given integral is to be extended for all positive values of the variables x, y and z such that

$$0 < x^2 + y^2 + z^2 < 1.$$

Now put $x^2 = u_1$ *i.e.*, $x = u_1^{1/2}$,

so that $dx = \frac{1}{2} u_1^{-1/2} du_1$,

$y^2 = u_2$ *i.e.*, $yu_2^{1/2}$,

so that $dy = \frac{1}{2} u_2^{-1/2} du_2$,

and $z^2 = u_3$ *i.e.*, $z = u_3^{1/2}$,

so that $dz = \frac{1}{2} u_3^{-1/2} du_3$.

With these substitutions the given condition reduces to

$$0 < u_1 + u_2 + u_3 < 1$$

and the required integral becomes

$$= \iiint \frac{\left(\frac{1}{2}\right)^3 u_1^{-1/2} u_2^{-1/2} u_3^{-1/2} du_1 du_2 dr_3}{\sqrt{(1 - u_1 - u_2 - u_3)}}, \text{ for } 0 < u_1 + u_2 + u_3 < 1$$

$$= \frac{1}{8} \iiint \frac{u_1^{1/2-1} u_2^{1/2-1} u_3^{1/2-1} du_1 du_2 dr_3}{\sqrt{\{1 - (u_1 - u_2 - u_3)\}}}$$

$$= \frac{1}{8} \frac{\left[\Gamma\left(\frac{1}{2}\right)\right]^3}{\Gamma\left(\frac{1}{2} + \frac{1}{2} + \frac{1}{2}\right)} \int_0^1 u^{3/2-1} \cdot \frac{1}{\sqrt{(1-u)}} du$$

(by Liouville's Extension of Dirichlet's Theorem)

$$= \frac{1}{8} \cdot \frac{[\sqrt{\pi}]^3}{\frac{1}{2}\sqrt{\pi}} \int_0^{\pi/2} \frac{\sin\theta . 2 \sin\theta \cos\theta \, d\theta}{\sqrt{(1 - \sin^2\theta)}}, \quad \text{putting } u = \sin^2\theta \text{ etc.}$$

$$= \frac{\pi}{2} \int_0^{\pi/2} \sin^2\theta \, d\theta = \frac{\pi}{2} \cdot \frac{1}{2} \cdot \frac{\pi}{2} = \frac{\pi^2}{8}.$$

Example 45:

Prove that

$$\iiint \frac{dx\,dy\,dz}{\sqrt{(a^2 - x^2 - y^2 - z^2)}} = \frac{\pi^2 a^2}{8}.$$

The integral being extended for all positive values of the variables for which the expression is real.

Solution:

The given expression is real when $x^2 + y^2 + z^2 < a^2$.

Therefore the required integral is to be extended to all positive values of x, y and z such that

$$0 < x^2 + y^2 + z^2 < a^2$$

i.e., $0 < x^2/a^2 + y^2/a^2 + z^2/a^2 < 1.$

Put $(x^2/a^2) = u_1$, $(y^2/a^2) = u_2$ and $(z^2/a^2) = u_3$

i.e., $x = au_1^{1/2}$, $y = au_2^{1/2}$ and $z = au_3^{1/2}$

so that $dx = \frac{1}{2} au_1^{-1/2} du_1$, $dy = \frac{1}{2} au_2^{-1/2} du_2$ and $dz = \frac{1}{2} au_3^{-1/2} du_3$.

Whit these substitutions the given condition reduces to

$$0 < u_1 + u_2 + u_3 < 1$$

and the required integral becomes

$$= \iiint \frac{\left(\frac{1}{2}\right)^3 .a^3 u_1^{-1/2} u_2^{-1/2} u_3^{-1/2} du_1 du_2 du_3}{a\sqrt{\{1-(u_1+u_2+u_3)\}}}$$

$$= \frac{a^2}{8} \iiint \frac{u_1^{1/2-1} u_2^{1/2-1} u_3^{1/2-1} du_1 du_2 du_3}{\sqrt{\{1-(u_1+u_2+u_3)\}}}$$

$$= \frac{a^2}{8} \cdot \frac{[\Gamma(1/2)]^3}{\Gamma\left(\frac{3}{2}\right)} \cdot \int_0^1 u^{3/2-1} \cdot \frac{1}{\sqrt{(1-u)}} du,$$

(by Liouville's extension of Dirichlet's theorem)

$$= \frac{a^2}{8} \cdot \frac{[\Gamma\pi]^3}{\frac{1}{2} \cdot \sqrt{\pi}} \cdot \int_0^{\pi/2} \frac{\sin\theta . 2\sin\theta\cos\theta\, d\theta}{\sqrt{(1-\sin^2\theta)}}$$

putting $u = \sin^2\theta$ etc.

$$= \frac{\pi a^2}{2} \int_0^{\pi/2} \sin^2\theta\, d\theta = \frac{\pi a^2}{2} \cdot \frac{1}{2} \cdot \frac{\pi}{2} = \frac{\pi^2 a^2}{8}.$$

Example 46:

Evaluate

$$\iiint_R (x+y+z+1)^2 \, dx\, dy\, dz,$$

where R is defined by $x \geq 0$, $y \geq 0$, $z \geq 0$, $x + y + z \leq 1$.

Solution:

As given x, y, z are all positive such that

$$0 \leq x + y + z \leq 1.$$

$$\therefore \iiint (x+y+z+1)^2 \, dx\, dy\, dz$$

$$= \iiint x^{1-1} y^{1-1} z^{1-1} \{(x+y+z)+1\}^2 \, dx\, dy\, dz$$

$$= \frac{\Gamma(1)\Gamma(1)\Gamma(1)}{\Gamma(1+1+1)} \int_0^1 (u+1)^2 . u^{1+1+1-1} du,$$

(by Liouville's extension of Dirichlet's theorem)

$$= \frac{1}{2}\int_0^1 (u^2 + 2u + 1) u^2 du = \frac{1}{2}\left[\frac{u^5}{5} + \frac{2u^4}{4} + \frac{u^3}{3}\right]_0^1$$

$$= \frac{1}{2}\left[\frac{1}{5} + \frac{1}{2} + \frac{1}{3}\right] = \frac{1}{2}\frac{(6+15+10)}{5 \times 2 \times 3} = \frac{1}{2} \cdot \frac{31}{30} = \frac{31}{60}.$$

Example 47:

Show that

$$\iiint \left(\frac{1 - x^2 - y^2}{1 + x^2 + y^2}\right)^{1/2} dx\, dy = \frac{\pi}{8}(\pi - 2)$$

over the positive quadrant of the circe $x^2 + y^2 = 1$.

Solution:

Here the given integral is to be extended to all positive values of x and y such that

$$0 \leq x^2 + y^2 \leq 1. \quad \text{...(1)}$$

Put $x^2 = u$, $y^2 = v$ *i.e.*, $x = u^{1/2}$, $y = v^{1/2}$ so that

$$dx = \frac{1}{2} u^{-1/2} du, \; dy = \frac{1}{2} v^{-1/2} dv.$$

With these substitutions the condition (1) becomes

$$0 \leq u + v \leq 1.$$

Hence the required integal

$$= \iint \left[\frac{1-(u+v)}{1+(u+v}\right]^{1/2} \frac{1}{4} u^{-1/2} v^{-1/2} du\, dv$$

$$= \frac{1}{4}\iint \left[\frac{1-(u+v)}{1+(u+v}\right]^{1/2} u^{(1/2)-1} v^{(1/2)-1} du\, dv,] \text{ (where } 0 \le u+v \le 1)$$

$$= \frac{1}{4}\frac{\Gamma(1/2)\Gamma(1/2)}{\Gamma\left(\frac{1}{2}+\frac{1}{2}\right)} \int_0^1 \left[\frac{1-h}{1+h}\right]^{1/2} h^{(1/2)+(1/2)-1} dh,$$

(by Liouville's extension of Dirichlet's theorem)

$$= \frac{1}{4}\frac{\sqrt{\pi}.\sqrt{\pi}}{\Gamma(1)} \int_0^1 \frac{1-h}{\sqrt{(1-h^2)}} dh$$

$$= \frac{\pi}{4}\int_0^{\pi/2} \frac{(1-\sin\theta)}{\cos\theta} \cos\theta\, d\theta,$$

putting $h = \sin\theta$ sot hat

$dh = \cos\theta\, d\theta$

$$= \frac{\pi}{4}[\theta + \cos\theta]_0^{\pi/2} = \frac{\pi}{4}\left[\frac{\pi}{2} - 1\right] = \frac{\pi}{8}(\pi - 2).$$

Example 48:

Prove that when x and y are postive and $x + y < h$,

$$\iint f'(x+y) x^{l-1} y^{-l} dx\, dy = \frac{\pi}{\sin l\pi}[f(h) - f(0)].$$

Solution:

The given integral

$$1 = \iint f'(x+y) x^{l-1} y^{(1-l)-1} dx\, dy, \text{ where } 0 < x+y < h$$

$$= \frac{\Gamma(l)\Gamma(1-l)}{\Gamma(l+1-l)} \int_0^h f'(u) u^{l+(1-l)-1} du$$

(by Liouvill's extension of Dirichlet's theorem)

$$= \frac{\Gamma(l)\Gamma(1-l)}{\Gamma(1)} \int_0^h f'(u) du$$

$$= \frac{\pi}{\sin \pi l}[f(u)]_0^h = \frac{\pi}{\sin \pi l}[f(h) - f(0)].$$

Example 49:

Evaluate $\iiint x^{\alpha} y^{\beta} z^{y} (1-x-y-z)^{\lambda}\, dx\, dy\, dz$ *over the interior of the tetrahedron formed by the coordinate planes and the plane $x + y + z = 1$.*

Solution:

Here the region of integration is bounded by the planes $x = 0$, $y = 0$ and $x + y + z = 1$. So the variables x, y, z take all positive values subject to the condition

$$0 < x + y + z < 1.$$

Hence the given integral

$$= \iiint x^{(\alpha+1)-1} y^{(\beta+1)-1} z^{(\gamma+1)-1} [1-(x+y+z)]^{\lambda}\, dx\, dy\, dz$$

$$= \frac{\Gamma(\alpha+1)\Gamma(\beta+1)\Gamma(\gamma+1)}{\Gamma(\alpha+\beta+\gamma+3)} \cdot$$

$$\int_0^1 u^{\alpha+1+\beta+1+\gamma+1-1} (1-u)^{\lambda}\, du,$$

(by Liouville's extension of Dirichlet's theorem)

$$= \frac{\Gamma(\alpha+1)\Gamma(\beta+1)\Gamma(\gamma+1)}{\Gamma(\alpha+\beta+\gamma+3)} \cdot$$

$$\int_0^1 u^{(\alpha+\beta+\gamma+3)-1} (1-u)^{(\lambda+1)-1}\, du$$

$$= \frac{\Gamma(\alpha+1)\Gamma(\beta+1)\Gamma(\gamma+1)}{\Gamma(\alpha+\beta+\gamma+3)} B(\alpha+\beta+\gamma+3.\lambda+1)$$

$$= \frac{\Gamma(\alpha+1)\Gamma(\beta+1)\Gamma(\gamma+1)}{\Gamma(\alpha+\beta+\gamma+3)} \cdot \frac{\Gamma(\alpha+\beta+\gamma+3)\Gamma(\lambda+1)}{\Gamma(\alpha+\beta+\gamma+\lambda+4)}$$

$$= \frac{\Gamma(\alpha+1)\Gamma(\beta+1)\Gamma(\gamma+1)\Gamma(\lambda+1)}{\Gamma(\alpha+\beta+\gamma+\lambda+4)}.$$

Example 50:

Evaluate

$\iiint \sqrt{(a^2b^2c^2 - b^2c^2x^2 - c^2a^2y^2 - a^2b^2z^2)}\, dx\, dy\, dz$ *taken throughout the ellipsoid* $x^2/a^2 + y^2/b^2 + z^2/c^2 = 1$.

Solution:

The given ellipsoid $x^2/a^2 + y^2/b^2 + z^2/c^2 = 1$ is symmetrical in all the eight octants Let us first evaluate the given integral over the region of the ellipsoid which lies in the positive octant *i.e.*, where x, , z are all positive.

Put $x^2/a^2 = u$, $y^2/b^2 = v$, $z^2/c^2 = w$.

Then $x = au^{1/2}$, $dx = \frac{1}{2} au^{-1/2} du$ etc.

Now the given integral extended over the positive octant of the given ellipsoid is

$$I = abc \iiint \sqrt{\left(1 - \frac{x^2}{a^2} - \frac{y^2}{b^2} - \frac{z^2}{c^2}\right)} dx\, dy\, dz,$$

where $0 < x^2/a^2 + y^2/b^2 + z^2/c^2 \leq 1$

$$= abc \iiint \sqrt{(1 - u - v - w)} . \frac{1}{8} abc\, u^{-1/2} v^{-1/2} w^{-1/2} du\, dv\, dw$$

where $0 < u + v + w \leq 1$

$$= \frac{a^2b^2c^2}{8} \iiint u^{(1/2)-1} v^{(1/2)-1} w^{(1/2)-1} \sqrt{\{1 - (u + v + w)\}} du\, dv\, dw$$

$$= \frac{a^2b^2c^2}{8} \cdot \frac{\left[\Gamma\left(\frac{1}{2}\right)\right]^3}{\Gamma(3/2)} \int_0^1 \sqrt{(1 - t)} . t^{1/2+1/2+1/2-1} dt,$$

by Liouvelle's extension of Dirichlet's theorem

$$= \frac{a^2b^2c^2}{8} \cdot \frac{(\sqrt{\pi})^3}{\frac{1}{2} . \sqrt{\pi}} \int_0^1 (1 - t)^{(3/2)-1} t^{(3/2)-1} dt$$

$$= \frac{a^2b^2c^2}{8} \cdot 2\pi \cdot \frac{\Gamma(3/2)\Gamma(3/2)}{\Gamma(3)} = \frac{\pi^2 a^2b^2c^2}{32}.$$

Hence if the integration is extended throughout the ellipsoid, the given integral $= 8I = 8 \cdot \dfrac{\pi^2 a^2b^2c^2}{32} = \dfrac{\pi^2 a^2b^2c^2}{4}$.

Example 51:

Evaluate $\iint \sqrt{\left\{\dfrac{1 - x^2/a^2 - y^2/b^2}{1 + x^2/a^2 + y^2/b^2}\right\}} dx\, dy$ *where* $x^2/a^2 + y^2/b^2 \leq 1$

Solution:

The ellipse $x^2/a^2 + y^2/b^2 = 1$ is symmetrical in all the four quadrants.

Let us first evaluate the given integral over the region of the ellipse $x^2/a^2 + y^2/b^2 = 1$ which lies in the first quadrant *i.e.*, where x and y are both positive.

Put $x^2/a^2 = u$, $y^2/b^2 = v$.

Then $x = au^{1/2}$, $dx = \dfrac{1}{2} au^{-1/2} du$,

$y = bv^{1/2}$, $dy = \dfrac{1}{2} bv^{-1/2} dv$.]

$\therefore$ the given integral extended over the region of the ellipse $x^2/a^2 + y^2/b^2 = 1$ which lies in the first quadrant is given by

$$I = \iint \sqrt{\left(\frac{1-u-v}{1+u+v}\right)} \cdot \frac{1}{2} abu^{-1/2} v^{-1/2} du\, dv \qquad \text{(where } 0 < u + v \le 1\text{)}$$

$$= \frac{ab}{4} \iint \sqrt{\left(\frac{1-(u-v)}{1+(u+v)}\right)} u^{(1/2)-1} v^{(1/2)-1} du\, dv$$

$$= \frac{ab}{4} \cdot \frac{\Gamma\left(\frac{1}{2}\right)\Gamma\left(\frac{1}{2}\right)}{\Gamma(1)} \int_0^1 \sqrt{\left(\frac{1-t}{1+t}\right)} . t^{1/2+1/2-1} dt$$

$$= \frac{\pi ab}{4} \cdot \int_0^1 \frac{1-t}{\sqrt{(1-t^2)}} dt = \frac{\pi ab}{4} \int_0^{\pi/2} \frac{1-\sin\theta}{\cos\theta} \cos\theta\, d\theta,$$

putting $t = \sin\theta$ so that $dt = \cos\theta\, d\theta$

$$= \frac{\pi ab}{4} \int_0^{\pi/2} (1 - \sin\theta) d\theta = \frac{\pi ab}{4} [\theta + \cos\theta]_0^{\pi/2}$$

$$= \frac{\pi ab}{4} \left[\left(\frac{\pi}{2} + 0\right) - (0+1)\right] = \frac{\pi ab}{4}\left(\frac{\pi}{2} - 1\right).$$

Hence the given integral extended over the whole region of the ellipse $x^2/a^2 + y^2/b^2 = 1$ 4. $I = \pi ab \left(\frac{1}{2}\pi - 1\right)$.

Example 52:

Prove that $I = \iiiint dx\, dy\, dz\, dw$, *for all positive values of the variables for which* $x^2 + y^2 + z^2 + w^2$ *is not less than a2 and not greater than* b^2 *is* $\pi^2(b^4 - a^4)/32$.

Solution:

We have to evaluate I subject to the condition

$a^2 < x^2 + y^2 + z^2 + w^2 < b^2$.

Putting $x^2 = u_1$ *i.e.*, $x = u_1^{1/2}$ dx . $\frac{1}{2} u^{-1/2} du_1$ etc, we get

$$I = \iiiint \frac{1}{2} u_1^{-1/2} . \frac{1}{2} u_2^{-1/2} . \frac{1}{2} u_3^{-1/2} . \frac{1}{2} u_4^{-1/2} du_1 du_2 du_3 du_4$$

subject to the condition $a^2 < u_1 + u_2 + u_3 + u_4 + < b^2$

$$\text{or } I = \frac{1}{16} \iiiint u_1^{(1/2)-1} u_2^{(1/2)-1} u_3^{(1/2)-1} u_4^{(1/2)-1} du_1 du_2 du_3 du_4$$

$$= \frac{1}{16} \cdot \frac{\left[\Gamma\left(\frac{1}{2}\right)\right]^4}{\Gamma\left(\frac{1}{2}+\frac{1}{2}+\frac{1}{2}+\frac{1}{2}\right)} \int_{a^2}^{b^2} t^{1/2+1/2+1/2+1/2-1} dt,$$

(by Liouville's theorem)

$$= \frac{(\sqrt{\pi})^4}{16\Gamma(2)} \int_{a^2}^{b^2} t\, dt = \frac{\pi^2}{16} \cdot \left[\frac{t^2}{2}\right]_{a^2}^{b^2} = \frac{\pi^2}{32}(b^4 - a^4).$$

Example 53:

Find the value of

$$\iiint xyz \sin(x+y+z) dx\, dy\, dz,$$

the integral being extended to all positive values of the variables subject to the condition $x + y + z \le \pi/2$.

Solution:

Here $0 < x + y + z \le \pi/2$.

$\therefore$ the required integral

$$I = \iiint \sin(x+y+z) x^{2-1} y^{2-1} z^{2-1} dx\, dy\, dz, \text{ where } 0 < x+y+z \le \pi/2$$

$$= \frac{\Gamma(2)\Gamma(2)\Gamma(2)}{\Gamma(2+2+2)} \int_0^{\pi/2} (\sin u) u^{2+2+2-1} du, \quad \text{(by Liouville's theorem)}$$

$$= \frac{1}{\Gamma(6)} \int_0^{\pi/2} u^5 \sin u\, du.$$

Applying successive integration by parts, we have

$$I = \frac{1}{5!}[u^5(-\cos u) - (5u^4)(-\sin u) + (20u^3)(\cos u)$$
$$-(60u^2)(\sin u) + (120u)(-\cos u) - 120(-\sin u)]_0^{\pi/2}.$$

In the above expression all the terms vanish for u = 0 and all those which involve cos u vanish for u = $\pi/2$.

$$\therefore I = \frac{1}{120}[-5(\pi/2)^4.(-1) - 60(\pi/2)^2.(1) - 120.(-1)]$$

$$= \frac{1}{120}\left[\frac{5\pi^4}{16} - 15\pi^2 + 120\right] = \frac{1}{384}[\pi^4 - 48\pi^2 + 384.$$

Example 54:

Evaluate $\iiint \sqrt{\left(\frac{1-x^2-y^2-z^2}{1+x^2+y^2+z^2}\right)} dx\,dy\,dz$ *integral being taken over all positive values of x, y, z such that*

$x^2 + y^2 + z^2 \le 1.$

Solution:

Put $x^2 = u,\ y^2 = v,\ z^2 = w.$

Then $x = u^{1/2},$

$$dx = \frac{1}{2}u^{-1/2}du;$$

$$y = v^{1/2},$$

$$dy = \frac{1}{2}v^{-1/2}dv;$$

$$z = w^{1/2},$$

$$dz = \frac{1}{2}w^{-1/2}dw.$$

$\therefore$ the given integral

$$= \frac{1}{8}\iiint \sqrt{\left\{\frac{1-(u+v+w)}{1+(u+v+w)}\right\}} u^{1/2-1} v^{1/2-1} w^{1/2-1} du\,dv\,dw,$$

where u, v, w are all +ive and $0 < u + v + w \le 1$

$$= \frac{1}{8} \cdot \frac{\left[\Gamma\left(\frac{1}{2}\right)\right]^3}{\Gamma\left(\frac{3}{2}\right)} \int_0^1 \sqrt{\left(\frac{1-t}{1+t}\right)} t^{1/2+1/2+1/2-1} dt$$

$$= \frac{\pi}{4}\int_0^1 \frac{1-t}{\sqrt{(1-t^2)}} t^{1/2} dt.$$

Put $t^2 = z$ or $t = z^{1/2}$ so that $dt = \frac{1}{2}z^{-1/2}dz.$

$$\therefore I = \frac{\pi}{4}\int_0^1 (1-z)^{-1/2}[1-z^{1/2}]z^{1/4} . \frac{1}{2}z^{-1/2}dz$$

$$= \frac{\pi}{4}\int_0^1 (1-z)^{-1/2}(z^{-1/4} - z^{1/4})dz$$

$$= \frac{\pi}{4}\int_0^1 [z^{(3/4)-1}(1-z)^{(1/2)-1} - z^{(5/4)-1}(1-z)^{(1/2)-1}]dz$$

$$= \frac{\pi}{4}\left[B\left(\frac{3}{4},\frac{1}{2}\right) - B\left(\frac{5}{4},\frac{1}{2}\right)\right].$$

Example 55:

Express the following integrals in terms of Beta function:

(i) $\int_0^1 x^m (1-x^2)^n dx, m > -1, n > -1;$

(ii) $\int_0^1 \frac{x^2}{\sqrt{(1-x^5)}} dx.$

Solution:

(i) We have

$$\int_0^1 x^m (1-x^2) dx = \int_0^1 x^{m-1}(1-x^2)^n . x\, dx$$

$$= \int_0^1 y^{(m-1)/2}(1-y)^n \cdot \frac{dy}{2},$$

putting $x^2 = y$ so that $2x\,dx = dy$

$$= \frac{1}{2}\int_0^1 y^{(m-1)/2}(1-y)^n dy = \frac{1}{2}\int_0^1 y^{[(m-1)/2]-1}(1-y)^{(n+1)-1} dy$$

$$= \frac{1}{2} B\left(\frac{1}{2}(m+1), n+1\right).$$

(ii) We have $\int_0^1 \frac{x^2}{\sqrt{(1-x^5)}} dx$

$$= \int_0^1 x^2 (1-x^5)^{-1/2} dx$$

$$= \int_0^1 x^2 \cdot \frac{1}{x^4}(1-x^5)^{-1/2} . x^4 dx = \int_0^1 x^{-2}(1-x^5)^{-1/2} x^4 dx$$

$$= \int_0^1 y^{2/5}(1-y)^{-1/2} \cdot \frac{1}{5} dy,$$ putting $x^5 = y$ so that $5x^4\,dx = dy$

$$= \frac{1}{5}\int_0^1 y^{-2/5}(1-y)^{-1/2}\, dy = \frac{1}{5}\int_0^1 y^{(3/5)-1}(1-y)^{(1/2)-1} dy$$

$$= \frac{1}{5} B\left(\frac{3}{5},\frac{1}{2}\right).$$

Example 57:

Prove that $\int_0^a (a-x)^{m-1}.x^{n-1}dx = a^{m+n-1}\,\mathbf{B}(m, n).$

Solution:

We have $\int_0^a (a-x)^{m-1}x^{n-1}dx$

$$= \int_0^1 (a-ay)^{m-1}(ay)^{n-1}a\,dy\text{, putting } x = ay$$

$$= \int_0^1 a^{(m-1)+(n-1)+1}(1-y)^{m-1}y^{n-1}dy$$

$$= a^{m+n-1}\int_0^1 y^{n-1}(1-y)^{m-1}dy$$

$$= a^{m+n-1}\,\mathbf{B}(n, m) = a^{m+n-1}\,\mathbf{B}(m, n), \qquad [\because \mathbf{B}(m, n) = \mathbf{B}(n, m)].$$

Example 58:

Show that if m, n are positive, then

$$\int_a^b (x-a)^{m-1}(b-x)^{n-1}dx = (b-a)^{m+n-1}.\,\mathbf{B}(m, n).$$

Solution:

The given integral is

$$\int_a^b (x-a)^{m-1}(b-x)^{n-1}dx$$

Put $x = a + (b-a)y$

so that $dx = (b-a)\,dy$.

Also when $x = a$,

$y =$ and when $x = b$, $y = 1$.

$$\therefore \int_a^b (x-a)^{m-1}(b-x)^{n-1}dx$$

$$= \int_0^1 [(b-a)y]^{m-1}[b-a-(b-a)y]^{n-1}.(b-a)dy$$

$$= \int_0^1 (b-a)^{m-1}.y^{m-1}.(b-a)^{n-1}.(1-y)^{n-1}.(b-a)dy$$

$$= (b-a)^{m+n-1}\int_0^1 y^{m-1}(1-y)^{n-1}dy$$

$$= (b-a)^{m+n-1}\,\mathbf{B}(m, n).$$

Example 59:

Show that

$$\int_0^1 \frac{x^{m-1}(1-x)^{n-1}}{(a+bx)^{m+n}}dx = \frac{1}{(a+b)^m . a^n} B(m, n).$$

Solution:

The given integral

$$I = \int_0^1 \frac{x^{m-1}(1-x)^{n-1}}{(a+bx)^{m+n}}dx$$

$$= \int_0^1 \left(\frac{x}{a+bx}\right)^{m-1} \cdot \left(\frac{1-x}{a+bx}\right)^{n-1} \cdot \frac{1}{(a+bx)^2}dx.$$

Put $\dfrac{x}{a+bx} = \dfrac{y}{a+b}$ so that $\dfrac{(a+bx).1 - x.b}{(a+bx)^2}dx = \dfrac{dy}{a+b}$

i.e., $$\frac{1}{(a+bx)^2}dx = \frac{dy}{a(a+b)}.$$

Further

$$\frac{1-x}{a+bx} = \frac{1}{a}\frac{a-ax}{a+bx} = \frac{1}{a}\left[\frac{a+bx-ax-bx}{a+bx}\right]$$

$$= \frac{1}{a}\left[1 - \frac{x(a+b)}{a+bx}\right] = \frac{1-y}{a}.$$

Also when x = 0, y = 0 and when x = 1, y = 1.

$$\therefore I = \int_0^1 \left(\frac{y}{a+b}\right)^{m-1} \left(\frac{1-y}{a}\right)^{n-1} \cdot \frac{dy}{a(a+b)}$$

$$= \frac{1}{(a+b)^m . a^n} \int_0^1 y^{m-1}(1-y)^{n-1}dy = \frac{B(m,n)}{(a+b)^m . a^n}.$$

Example 60:

Show that

$$\int_0^1 \frac{x^{m-1}(1-x)^{n-1}}{(a+x)^{m+n}}dx = \frac{B(m,n)}{a^n(1+a)^m}.$$

Solution:

Do your self.

Here we have b = 1.

Example 61:

Show that

$$\int_0^{\pi/2} \sin^p \theta \cos^q \theta d\theta = \frac{1}{2} B\left(\frac{p+1}{2}, \frac{q+1}{2}\right), p > -1, q > -1.$$

Deduce that $\int_0^2 x^4 (8 - x^3)^{-1/3} dx = \frac{16}{3} B\left(\frac{5}{3}, \frac{2}{3}\right)$.

Solution:

We have $\int_0^{\pi/2} \sin^p \theta \cos^q \theta d\theta$

$$= \int_0^{\pi/2} \sin^{p-1} \theta \cos^{q-1} \theta \sin\theta \cos\theta d\theta$$

$$= \int_0^{\pi/2} \sin^{p-1} \theta . (1 - \sin^2 \theta)^{\frac{1}{2}(q-1)} \sin\theta \cos\theta d\theta.$$

Put $\sin^2\theta = x$

so that $2 \sin \theta \cos \theta \, d\theta = dx$

Also when $\theta = 0, x = 0$

and when $\theta = \frac{1}{2}\pi, x = 1.$

$\therefore$ the given integral

$$= \int_0^1 x^{(p-1)/2} . (1 - x)^{(q-1)/2} . \frac{1}{2} dx$$

$$= \frac{1}{2} \int_0^1 x^{\{(p+1)/2\}-1} (1 - x)^{\{(q+1)/2\}-1} dx$$

$$= 12 B\left(\frac{p+1}{2}, \frac{q+1}{2}\right), \quad \text{...(1)}$$

where $\frac{p+1}{2} > 0$ *i.e.*, $p > -1$

and $\frac{q+1}{2} > 0$ *i.e.*, $q > -1$.

Second part

We have

$$I = \int_0^2 x^4 (8 - x^3)^{-1/3} dx = \int_0^2 x^2 (8 - x^3)^{-1/3} . x^2 dx.$$

Put $x^3 = 8 \sin^2\theta$

so that $3x^2 \, dx = 16 \sin \theta \cos \theta \, d\theta$.

Also when $x = 0, \theta = 0$

and when $x = 2, \theta = \frac{1}{2}\pi$.

$$\therefore\ I = \int_0^{\pi/2} (8\sin^2\theta)^{2/3}(8 - 8\sin^2\theta)^{-1/3} . \frac{16}{3}\sin\theta\cos\theta\, d\theta$$

$$= \int_0^{\pi/2} 4\sin^{4/3}\theta . \frac{1}{2}\cos^{-2/3}\theta \cdot \frac{16}{3}\sin\theta\cos\theta\, d\theta$$

$$= \frac{32}{3}\int_0^{\pi/2} \sin^{7/3}\theta\cos^{1/3}\theta\, d\theta$$

$$= \frac{32}{3}\cdot\frac{1}{2}B\left(\frac{\frac{7}{3}+1}{2}, \frac{\frac{1}{3}+1}{2}\right), \qquad \left[\text{from}(1); \text{here } p = \frac{7}{3}, q = \frac{1}{3}\right]$$

$$= \frac{16}{3}B\left(\frac{5}{3}, \frac{2}{3}\right).$$

Example 62:

Prove that if p, q are positive, then

(i) $\dfrac{B(p, q+1)}{q} = \dfrac{B(p+1, q)}{p} = \dfrac{B(p, q)}{p+q}$;

(ii) $\mathbf{B}(p, q) = \mathbf{B}(p + 1, q) + \mathbf{B}(p, q + 1)$.

Solution:

(i) We have $\dfrac{B(p, q+1)}{q} = \dfrac{1}{q}B(p, q+1)$

$$= \frac{1}{q}\int_0^2 x^{p-1}(1-x)^{(q+1)-1}\,dx = \frac{1}{q}\int_0^1 (1-x)^q . x^{p-1}\,dx$$

$$= \frac{1}{q}\left[\left\{(1-x)^q . \frac{x^p}{p}\right\}_0^1 - \int_0^1 q(1-x)^{q-1}(-1)\cdot\frac{x^p}{p}\,dx\right],$$

(integrating by parts)

$$= \frac{1}{q}\left[0 + \frac{q}{p}\int_0^1 x^{(p+1)-1}(1-x)^{q-1}\,dx\right] = \frac{1}{p}B(p+1, q). \qquad ...(1)$$

Again $\dfrac{B(p, q+1)}{q} = \dfrac{1}{p}B(p+1, q)$, from (1)

$$= \frac{1}{p}\int_0^1 x^p(1-x)^{q-1}\,dx, \qquad \text{(by the def. of Beta function)}$$

$$= \frac{1}{p}\int_0^1 x^{p-1} . x(1-x)^{q-1}\,dx$$

$$= \frac{1}{p}\int_0^1 x^{p-1}[1-(1-x)](1-x)^{q-1}\,dx$$

$$= \frac{1}{p}\int_0^1 x^{p-1}(1-x)^{q-1}\,dx - \frac{1}{p}\int_0^1 x^{p-1}(1-x)^{q}\,dx$$

$$= \frac{1}{p}B(p,q) - \frac{1}{p}B(p,q+1).$$

$$\therefore \frac{B(p,q+1)}{q} + \frac{B(p,q+1)}{p} = \frac{B(p,q)}{p}$$

$$\text{or } \frac{B(p,q+1)}{q} = \frac{B(p,q)}{p+q}. \qquad ...(2)$$

Now from (1) and (2), we have

$$\frac{B(p,q+1)}{q} = \frac{B(p+1,q)}{p} = \frac{B(p,q)}{p+q}. \qquad ...(3)$$

From (3), we have

$$\mathbf{B}(p, q+1) = \frac{q}{p+q}\mathbf{B}(p, q)$$

and $\mathbf{B}(p+1, q) = \dfrac{p}{p+q}B(p,q)$.

Adding, we get

$\mathbf{B}(p+1, q) + \mathbf{B}(p, q+1)$

$$= \frac{p}{p+q}B(p,q) + \frac{q}{p+q}B(p,q) = \frac{p+q}{p+q}B(p,q) = \mathbf{B}(p, q).$$

Example 63:

Prove that

$$\frac{B(m+1,n)}{B(m,n)} = \frac{m}{m+n}.$$

Solution:

We have

$\mathbf{B}(m+1, n) = \mathbf{B}(n, m+1)$ (by the symmetry of Beta function)

$$= \int_0^1 x^{n-1}(1-x)^{(m+1)-1}\,dx = \int_0^1 (1-x)^m x^{n-1}dx$$

$$= \left[(1-x)^m \cdot \frac{x^n}{n}\right]_0^1 - \int_0^1 m(1-x)^{m-1}(-1)\cdot\frac{x^n}{n}\,dx,$$

(integrating by parts)

$$= 0 + \frac{m}{n}\int_0^1 x^{n-1}.x(1-x)^{m-1}\,dx$$

$$= \frac{m}{n}\int_0^1 x^{n-1}[1-(1-x)](1-x)^{m-1}\,dx$$

$$= \frac{m}{n}\left[\int_0^1 x^{n-1}(1-x)^{m-1}\,dx - \int_0^1 x^{n-1}(1-x)^m\,dx\right]$$

$$= \frac{m}{n}[B(n,m) - B(n,m+1)] = \frac{m}{n}B(m,n) - \frac{m}{n}B(m+1,n)$$

or $\left(1+\frac{m}{n}\right)B(m+1,n) = \frac{m}{n}B(m,n)$

or $(n + m)\, \mathbf{B}\,(m + 1, n) = m\, \mathbf{B}\,(m, n)$

or $\dfrac{B(m+1,n)}{B(m,n)} = \dfrac{m}{m+n}.$

Example 64:

Prove that

$$\int_0^\infty \frac{x^{m-1} - x^{n-1}}{(1-x)^{m+n}}\,dx = 0,\ m > 0,\ n > 0.$$

Solution:

The given integral is

$$= \int_0^\infty \frac{x^{m-1}}{(1+x)^{m+n}}\,dx - \int_0^\infty \frac{x^{n-1}}{(1+x)^{m+n}}\,dx$$

$= \mathbf{B}\,(m, n) - \mathbf{B}\,(n, m) = \mathbf{B}\,(m, n) - \mathbf{B}\,(m, n) = 0.$

Example 65:

Express $\int_0^\infty \frac{x^{m-1}}{(a+bx)^{m+n}}\,dx$ *in terms of Beta function where* $m > 0$, $n > 0$, $a > 0$, $b > 0$.

Solution:

In the given integral put $bx = ay$ *i.e.*, $x = (a/b)\, y$

so that $dx = (a/b)\, dy$.

When $x = 0, y = 0$

and when $x \to \infty, y \to \infty$.

$$\therefore \int_0^\infty \frac{x^{m-1}}{(a+bx)^{m+n}}dx = \int_0^\infty \left(\frac{a}{b}y\right)^{m-1} \cdot \frac{1}{(a+ay)^{m+n}} \cdot \frac{a}{b}dy$$

$$= \int_0^\infty \frac{a^{m-1}y^{m-1}a}{b^{m-1}.a^{m+n}(1+y)^{m+n}\,b}dy = \frac{1}{a^n b^m}\int_0^\infty \frac{y^{m-1}}{(1+y)^{m+n}}dy$$

$$= \frac{1}{a^n b^m}B(m,n), \text{ [by * 3].}$$

Example 66:

Evaluate the following integrals.

(i) $\int_0^\infty \frac{x^8(1-x^6)}{(1+x)^{24}}dx$,

(ii) $\int_0^\infty \frac{x^4(1-x^5)}{(1+x)^{15}}dx$,

(iii) $\int_0^\infty \frac{x\,dx}{1+x^6}$

Solution:

(i) We have

$$\int_0^\infty \frac{x^8(1-x^6)}{(1+x)^{24}}dx = \int_0^\infty \frac{x^8dx}{(1+x)^{24}} - \int_0^\infty \frac{x^{14}}{(1+x)^{24}}dx$$

$$= \int_0^\infty \frac{x^{9-1}}{(1+x)^{9+15}}dx - \int_0^\infty \frac{x^{15-1}}{(1+x)^{15+9}}dx$$

$= \mathbf{B}(9, 15) - \mathbf{B}(15, 9)$,

$= \mathbf{B}(9, 15) - \mathbf{B}(9, 15)$, (by symmetry of Beta function)

$= 0$.

(ii) We have

$$\int_0^\infty \frac{x^4(1+x^5)}{(1+x)^{15}}dx = \int_0^\infty \frac{x^4dx}{(1+x)^{15}} + \int_0^\infty \frac{x^9dx}{(1+x)^{15}}$$

$$= \int_0^\infty \frac{x^{5-1}}{(1+x)^{5+10}}dx + \int_0^\infty \frac{x^{10-1}}{(1+x)^{10+5}}dx$$

$= B(5, 10) + B(10, 5) = B(5, 10) + B(5, 10)$

$$= 2B\,(5, 10) = 2\,\frac{\Gamma 5\Gamma 10}{\Gamma 15}$$

$$= 2\cdot\frac{4.3.2.1}{14.13.12.11.10} = \frac{1}{5005}.$$

(iii) Let $I = \int_0^\infty \frac{x\,dx}{1+x^6}$.

Put $x^6 = y$ or $x = y^{1/6}$,

so that $dx = \frac{1}{6}\,y^{-5/6}\,dy$.

$$\therefore I = \frac{1}{6}\int_0^\infty \frac{y^{1/6}y^{-5/6}}{1+y}dy = \frac{1}{6}\int_0^\infty \frac{y^{-2/3}}{1+y}$$

$$= \frac{1}{6}\int_0^\infty \frac{y^{(1/3)-1}}{(1+y)^{(1/3)+(2/3)}}\,dy = \frac{1}{6}\,\mathbf{B}\left(\frac{1}{3},\frac{2}{3}\right),$$

$$= \frac{1}{6}\frac{\Gamma\frac{1}{3}\Gamma\frac{2}{3}}{\Gamma\left(\frac{1}{3}+\frac{2}{3}\right)} = \frac{1}{6}\frac{\Gamma\frac{1}{3}\Gamma\left(1-\frac{1}{3}\right)}{\Gamma 1}$$

$$= \frac{1}{6}\cdot\frac{\pi}{\sin\frac{1}{3}\pi}, \qquad \left[\because \Gamma n\Gamma(1-n) = \frac{\pi}{\sin n\pi}\right]$$

$$= \frac{1}{6}\cdot\frac{\pi}{(\sqrt{3}/2)} = \frac{1}{6}\cdot\frac{2\pi}{\sqrt{3}} = \frac{\pi}{3\sqrt{3}}.$$

Example 67:

Show that

$$\int_0^1 \frac{dx}{(1-x^n)^{1/2}} = \frac{\sqrt{\pi}\,\Gamma(1/n)}{n\Gamma(1/n+1/2)}$$

Solution:

Let $x^n = \sin^2\theta$

i.e., $x = \sin^{2/n}\theta$

so that $dx = \frac{2}{n}\,\sin{(2/n)^{-1}}\,\theta\,\cos\theta\,d\theta$.

Then $\int_0^1 \frac{dx}{\sqrt{(1-x^n)}} = \frac{2}{n}\int_0^{\pi/2} \frac{\sin^{(2/n)-1}\theta\cos\theta\,d\theta}{\cos\theta}$

$$= \frac{2}{n}\int_0^{\pi/2} \sin^{(2/n)-1}\theta \cos^0\theta \, d\theta$$

$$= \frac{2}{n} \cdot \frac{\Gamma(1/n)\Gamma\left(\frac{1}{2}\right)}{2\Gamma(1/n+1/2)} = \frac{\sqrt{\pi}}{n} \cdot \frac{\Gamma(1/n)}{\Gamma\left(1/n+\frac{1}{2}\right)}.$$

Example 68:

Evaluate $\int_0^1 \frac{x^{m-1}+x^{n-1}}{(1+x)^{m+n}} dx.$

Solution:

We have

$$\int_0^1 \frac{x^{m-1}+x^{n-1}}{(1+x)^{m+n}} dx = \int_0^1 \frac{x^{m-1}dx}{(1+x)^{m+n}} + \int_0^1 \frac{x^{n-1}dx}{(1+x)^{m+n}}.$$

Now in the second integral on the R.H.S. of (1), we put $x = 1/y$ so that $dx = -(1/y^2)\,dy$; also when $x \to 0$, $y \to \infty$ and when $x = 1$, $y = 1$.

$$\therefore \int_0^1 \frac{x^{n-1}dx}{(1+x)^{m+n}} = \int_\infty^1 \frac{(1/y)^{n-1}}{(1+1/y)^{m+n}}\left(-\frac{1}{y^2}dy\right)$$

$$= -\int_\infty^1 \frac{y^{m+n}dy}{(1+y)^{m+n}.y^{n-1}.y^2} = \int_1^\infty \frac{y^{m-1}}{(1+y)^{m+n}} dy$$

$$= \int_1^\infty \frac{x^{m-1}}{(1+x)^{m+n}} dx, \qquad \left[\because \int_a^b f(x)dx = \int_a^b f(y)dy\right].$$

Now from (1), we have

$$\int_0^1 \frac{x^{m-1}+x^{n-1}}{(1+x)^{m+n}} dx = \int_0^1 \frac{x^{m-1}}{(1+x)^{m+n}} dx + \int_1^\infty \frac{x^{m-1}}{(1+x)^{m+n}} dx$$

$= \int_0^\infty \frac{x^{m-1}}{(1+x)^{m+n}} dx$, by a property of definite integrals

$$= B(m, n) = \frac{\Gamma(m)\Gamma(n)}{\Gamma(m+n)}.$$

Example 69:

Prove that

$$B(m, n) = \int_0^1 \frac{x^{m-1}+x^{n-1}}{(1+x)^{m+n}} dx.$$

Example 70:

Show that

$$\int_0^{\pi/2} (\tan x)^n\, dx = \frac{\pi}{2}\sec\frac{n\pi}{2}, \text{ where } -1 < n < 1.$$

Solution:

We have

$$\int_0^{\pi/2} \tan^n x\, dx = \int_0^{\pi/2} \frac{\sin^n x}{\cos^n x} dx = \int_0^{\pi/2} \sin^n x \cos^{-n} x\, dx$$

$$= \frac{\Gamma\frac{1}{2}(n+1).\Gamma\frac{1}{2}(-n+1)}{2\Gamma\frac{1}{2}(n-n+2)},$$

where $-n+1>0$ *i.e.*, $n<1$

and $n+1>0$ *i.e.*, $n>-1$

$$= \frac{1}{2}\Gamma\frac{1}{2}(n+1)\Gamma\frac{1}{2}(1-n) = \frac{1}{2}\Gamma\frac{1}{2}(n+1)\Gamma\left[1-\frac{1}{2}(n+1)\right]$$

$$= \frac{\frac{1}{2}\pi}{\sin\frac{1}{2}(n+1)\pi},$$

$$= \frac{\pi}{2}\cdot\frac{1}{\sin\left(\frac{1}{2}\pi+\frac{1}{2}n\pi\right)} = \frac{\pi}{2}\cdot\frac{1}{\cos\left(\frac{1}{2}n\pi\right)} = \frac{\pi}{2}\sec\frac{n\pi}{2},$$

where $-1<n<1$.

Example 71:

Show that

$$\int_0^{\pi/2} \sqrt{(\sin\theta)}d\theta \cdot \int_0^{\theta/2} \frac{d\theta}{\sqrt{(\sin\theta)}} = \pi.$$

Solution:

The given integral is

$$= \int_0^{\pi/2} \sin^{-1/2}\theta\, d\theta \times \int_0^{\pi/2} \sin^{1/2}\theta\, d\theta$$

$$= \int_0^{\pi/2} \sin^{-1/2}\theta \cos^0\theta\, d\theta \times \int_0^{\pi/2} \sin^{1/2}\theta\cos^0\theta\, d\theta$$

$$= \frac{\Gamma\left\{\frac{1}{2}\left(-\frac{1}{2}+1\right)\right\}\Gamma\left\{\frac{1}{2}(0+1)\right\}}{2\Gamma\left\{\frac{1}{2}\left(-\frac{1}{2}+0+2\right)\right\}} \times \frac{\Gamma\left\{\frac{1}{2}\left(\frac{1}{2}+1\right)\right\}\Gamma\left\{\frac{1}{2}(0+1)\right\}}{2\Gamma\left\{\frac{1}{2}\left(\frac{1}{2}+0+2\right)\right\}}$$

$$= \frac{\Gamma\left(\frac{1}{4}\right)\Gamma\left(\frac{1}{2}\right)}{2\Gamma\left(\frac{3}{4}\right)} \times \frac{\Gamma\left(\frac{3}{4}\right)\Gamma\left(\frac{1}{2}\right)}{2.\Gamma\left(\frac{5}{4}\right)} = \frac{\Gamma\frac{1}{4}.\sqrt{\pi}.\sqrt{\pi}}{2.2.\frac{1}{4}\Gamma\left(\frac{1}{4}\right)} = \pi$$

Example 72:

Prove that

(i) $\int_0^{\infty} e^{-ax} x^{n-1} dx = \frac{\Gamma(n)}{a^n}$;

(ii) $\int_0^{\infty} x^{2n-1} e^{-ax^2} dx = \frac{\Gamma(n)}{2a^n}$;

(iii) $\int_0^1 \left(\log\frac{1}{2}\right)^{n-1} dx = \Gamma(n)$;

(iv) $\int_0^{\infty} x^m e^{-ax^n} dx = \frac{\Gamma[(m+1)/n]}{na^{(m+1)/n}}$;

(v) $\int_0^1 \frac{dx}{\sqrt{(-\log x)}} = \sqrt{\pi}$.

(vi) $\Gamma(m)\,\Gamma(n) = \Gamma(m+n) \int_0^{\infty} \frac{x^{n-1}dx}{(1+x)^{m+n}}$.

Solution:

(i) Let $I = \int_0^{\infty} e^{-ax} x^{n-1} dx$.

Put $ax = y$

so that $a\,dx = dy$.

When $x = 0$, $y = 0$

and when $x \to \infty$, $y \to \infty$.

$$\therefore I = \int_0^{\infty} e^{-y}\left(\frac{y}{a}\right)^{n-1} \cdot \frac{1}{a} dy = \frac{1}{a^n}\int_0^{\infty} e^{-y} y^{n-1} dy$$

$= \frac{1}{a^n}\Gamma(n)$, by the definition of Gamma function.

(ii) Let $I = \int_0^\infty x^{2n-1} e^{-ax^2} dx = \int_0^\infty x^{2n-2} e^{-ax^2} x\, dx$.

Put $ax^2 = z$

so that $2ax\, dx = dz = dz$.

When $x = O$, $z = 0$

and when $x \to \infty$, $z \to \infty$.

$$I = \int_0^\infty \left(\frac{z}{a}\right)^{n-1} e^{-z} \frac{1}{2a} dz = \frac{1}{2a^n} \int_0^\infty e^{-z} z^{n-1} dz$$

$$= \frac{1}{2a^n} \Gamma(n), \text{ by definition of Gamma function.}$$

(iii) Let $I = \int_0^1 \left(\log \frac{1}{x}\right)^{n-1} dx$.

Put $\log(1/x) = y$ *i.e.*, $1/x = e^y$ *i.e.*, $x = e^{-y}$ so that $dx = -e^{-y} dy$.

Also when $x \to 0$, $y \to \infty$ and when $x = 1$, $y = 0$. [Note that $\log \infty = \infty$]

$$\therefore I = -\int_\infty^0 y^{n-1} e^{-y} dy = \int_0^\infty e^{-y} y^{n-1} dy = \Gamma(n).$$

(iv) Let $I = \int_0^\infty x^m e^{-ax^n} dx = \int_0^\infty \frac{x^m}{x^{n-1}} e^{-ax^n} x^{n-1} dx$

$$= \int_0^\infty x^{m-n+1} e^{-ax^n} x^{n-1} dx.$$

Put $ax^n = t$

so that $an\, x^{n-1}\, dx = dt$.

Also when $x = 0$, $t = 0$

and when $x \to \infty$, $t \to \infty$.

$$\therefore I = \int_0^\infty \left(\frac{t}{a}\right)^{(m-n+1)/n} e^{-t} \cdot \frac{1}{na} dt, \qquad \left[\because ax^n = t \Rightarrow x = \left(\frac{t}{a}\right)^{1/n}\right]$$

$$= \frac{1}{na.a^{(m-n+1)/n}} \int_0^\infty t^{\{(m+1)/n\}-1} e^{-t} dt$$

$$= \frac{1}{na^{(m+1)/n}} \Gamma\{(m+1)/n\}, \quad \text{(by the definition of Gamma function)}$$

(v) Let $I = \int_0^1 \frac{dx}{\sqrt{(-\log x)}} = \int_0^1 \frac{dx}{\sqrt{\{\log(1/x)\}}} = \int_0^1 \left(\log \frac{1}{x}\right)^{-1/2} dx$.

Put $\log(1/x) = y$ *i.e.*, $1/x = e^y$

i.e., x e^{-y} so that dx = – e^{-y} dy.

Also when x → ∞, y → ∞

and when x = 1, y = 0.

$$\therefore I = -\int_{\infty}^{0} y^{-1/2} e^{-y} dy = \int_{0}^{\infty} e^{-y} y^{1/2-1} dy = \Gamma\left(\frac{1}{2}\right).$$

(by the der. of Gamma function)

$$= \sqrt{\pi}.$$

(vi) We have

$$\frac{\Gamma(n)}{z^n} = \int_0^{\infty} e^{-zx} x^{n-1} dx.$$

$$\therefore \Gamma(n) = z^n \int_0^{\infty} e^{-zx} . x^{m-1} dx = \int_0^{\infty} z^n e^{-zx} x^{n-1} dx.$$

Multiplying both sides $e^{-z} z^{m-1}$, we get

$$\Gamma(n)\, e^{-z} z^{m-1} = \int_0^{\infty} e^{-z(1+x)} z^{m+n-1} x^{m-1} dx.$$

Now proceed as in 7.

Example 73:

Show that

$$\int_0^1 \frac{x^2 dx}{(1+x^4)^{1/2}} \times \int_0^1 \frac{dx}{(1+x^4)^{1/2}} = \frac{\pi}{4\sqrt{2}}.$$

Solution:

In the first integral put x^2 = sin θ sot at 2x dx = cos θ dθ and the corresponding limits for θ are 0 to θ/2.

$$\therefore \int_0^1 \frac{x^2 dx}{(1-x^4)^{1/2}} = \int_0^1 \frac{\frac{1}{2} x . 2x\, dx}{(1-x^4)^{1/2}} = \frac{1}{2}\int_0^{\pi/2} \frac{\sin^{1/2}\theta . \cos\theta\, d\theta}{(1-\sin^2\theta)^{1/2}}$$

$$= \frac{1}{2}\int_0^{\pi/2} \frac{\sqrt{(\sin\theta)} . \cos\theta\, d\theta}{\cos\theta} = \frac{1}{2}\int_0^{\pi/2} \sqrt{(\sin\theta)} d\theta.$$

Again, i 2nd integral put x^2 = tan θ so that

$$2x\, dx = \sec^2\theta\, d\theta$$

and the corresponding limits for θ are 0 to π/4.

$$\therefore \int_0^1 \frac{dx}{(1+x^4)^{1/2}} = \frac{1}{2}\int_0^{\pi/2} \frac{\sec^2\theta\, d\theta}{\sqrt{(\tan\theta)} \sec\theta}$$

$$= \frac{1}{2}\int_0^{\pi/4} \frac{d\theta}{\sqrt{(\sin\theta\cos\theta)}}$$

$$= \frac{1}{\sqrt{2}}\int_0^{\pi/4} \frac{d\theta}{\sqrt{(\sin 2\theta)}} = \frac{1}{2\sqrt{2}}\int_0^{\pi/2} \frac{1}{\sqrt{(\sin\phi)}}d\phi, \text{ where } 2\theta = \phi.$$

Hence the given integral becomes

$$\int_0^1 \frac{x^2 dx}{(1-x^4)^{1/2}} \times \int_0^1 \frac{dx}{(1+x^4)^{12}}$$

$$= \frac{1}{2}\int_0^{\pi/2} \sqrt{(\sin\theta)}d\theta \times \frac{1}{2\sqrt{2}}\int_0^{\pi/2} \frac{d\phi}{\sqrt{(\sin\phi)}}$$

$$= \frac{1}{4\sqrt{2}} \cdot \int_0^{\pi/2} \sqrt{(\sin\theta)}d\theta \times \int_0^{\pi/2} \frac{d\theta}{\sqrt{(\sin\theta)}}$$

$$= \frac{1}{4\sqrt{2}} \cdot \pi$$

Example 74:

Show that

$$\int_0^a \frac{dx}{(a^n - x^n)^{1/n}} = \frac{1}{n} \cdot \frac{\pi}{\sin(\pi/n)}.$$

Solution:

In the given integral put $x^n = a^n \sin^2\theta$

i.e., $x = a \sin^{2/n}\theta\, d\theta$

so that $dx = \sin^{(2/n)} \sin^{(2/n)-1}\theta \cos\theta\, d\theta$.

Also when $x = 0$, $\theta = 0$

and when $x = a$, $\theta = \pi/2$.

$$\therefore \int_0^a \frac{dx}{(a^n - x^n)^{1/n}} = \frac{2a}{n}\int_0^{\pi/2} \frac{\sin^{(2/n)-1}\theta\cos\theta\, d\theta}{a\cos^{2/n}\theta}$$

$$= \frac{2}{n}\int_0^{\pi/2} \sin^{(2/n)-1}\theta\cos^{1-(2/n)}\theta d\theta = \frac{\frac{2}{n}\Gamma\frac{1}{n}\Gamma\left(1-\frac{1}{n}\right)}{2\Gamma 1}$$

$$= \frac{1}{n}\cdot\Gamma\frac{1}{n}\Gamma\left(1-\frac{1}{n}\right) = \frac{1}{n}\cdot\frac{\pi}{\sin(\pi/n)}. \qquad \left[\Gamma(n)\Gamma(1-n) = \frac{\pi}{\sin n\pi}\right].$$

Example 75:

Show that

$$\Gamma\left(\frac{1}{4}\right)\Gamma\left(\frac{3}{4}\right) = 2\int_0^{\pi/2} \sqrt{(\tan\theta)}d\theta = 4\int_0^{\infty} \frac{x^2 dx}{1+x^4} = \pi\sqrt{2}.$$

Solution:

Let $I = 4\int_0^{\infty} \frac{x^2 dx}{1+x^4} = 4\int_0^{\infty} \frac{x.x\,dx}{1+x^4}$.

Put $x^2 = \tan\theta$

so that $2x\,dx = \sec^2\theta\,d\theta$.

Also when $x = 0$, $\theta = 0$

and when $x \to \infty$, $\theta \to \pi/2$.

$$\text{Then } I = 4\int_0^{\pi/2} \frac{\sqrt{(\tan\theta)}\cdot\frac{1}{2}\sec^2\theta\,d\theta}{(1+\tan^2\theta)} = 2\int_0^{\pi/2} \sqrt{(\tan\theta)}d\theta \qquad ...(1)$$

$$= 2\int_0^{\pi/2} \sin^{1/2}\theta\cos^{-1/2}\theta\,d\theta$$

$$= 2\frac{\Gamma\left\{\frac{1}{2}\left(\frac{1}{2}+1\right)\right\}\Gamma\left\{\frac{1}{2}\left(-\frac{1}{2}+1\right)\right\}}{2\Gamma\left\{\frac{1}{2}\left(\frac{1}{2}-\frac{1}{2}+2\right)\right\}} = \frac{2\Gamma\left(\frac{3}{4}\right)\Gamma\left(\frac{1}{2}\right)}{2\Gamma 1}$$

$$= \Gamma\left(\frac{1}{4}\right)\Gamma\left(\frac{3}{4}\right) \qquad ...(2)$$

$$= \Gamma\left(\frac{1}{4}\right)\Gamma\left(1-\frac{1}{4}\right) = \frac{\pi}{\sin\frac{1}{4}\pi}, \qquad \left[\because \Gamma(n)\Gamma(1-n) = \frac{\pi}{\sin n\pi}\right]$$

$$= \frac{\pi}{1/\sqrt{2}}\ \pi\sqrt{2}. \qquad ...(3)$$

From (1), (2) and (3), the required result follows.

3

Convergence of Improper Integrals

3.1 Some Definitions

1. *Infinite Interval* : The interval whose length (range) is infinite is said to be an *infinite interval*. Thus, the intervals (a, ∞), $(-\infty, b)$ and $(-\infty, \infty)$ are infinite intervals.

2. *Bounded Functions* : A function f (x) is said to be *bounded* over the interval I if there exist two real numbers a and b (b > a) such that

$$a \leq f(x) \leq b \text{ for all } x \in I.$$

A function f (x) is said to be unbounded at a point, if it becomes infinite at that point. Thus the function

$$f(x) = x/\{(x-1)(x-2)\}$$

is unbounded each of the points x = 1 and x = 2.

3. *Monotonic Functions* : A real valued function f defined on an interval I is said to be *monotonically* increasing if

$$x > y \Rightarrow f(x) > f(x) \ \forall\ x, y \in I$$

and *monotonically decreasing* if

$$x > y \Rightarrow f(x) < f(x) \ \forall\ x, y \in I.$$

A function f defined on an interval I is said to be a monotonic function if it is either monotonically decreasing or monotonically increasing on I.

For example the function f defined by f (x) = sin x is monotonically increasing in the interval $0 \leq x \leq \frac{1}{2}\pi$ and monotonically decreasing in the interval $\frac{1}{2}\pi \leq x \leq \pi$.

4. *Proper Integral* : The definite integral $\int_a^b f(x)\,dx$ is said to be a *proper integral* if the range of integration is finite and the integrand f (x) is bounded. The integral $\int_0^{\pi/2} \sin x\,dx$ is a proper integral.

Also $\int_0^1 \frac{\sin x}{x} dx$ is an example of a proper integral because

$$\lim_{x \to 0} \frac{\sin x}{x} = 1.$$

5. *Improper Integral* : The definite integral $\int_a^b f(x)\,dx$ is said to be an *improper integral* if *(i)* the interval (a, b) is not finite (*i.e.*, is infinite) and the function f (x) is bounded over this interval; or *(ii)* the interval (a, b) is finite and f (x) is not bounded over this interval; or *(iii)* neither the interval (a, b) is finite nor f (x) is bounded over it.

6. *Improper Integrals of the First Kind or Infinite Integrals* : A definite integral $\int_a^b f(x)dx$ in which the range of integration is infinite (*i.e.*, either b = ∞ or a = – ∞ or both) and the integrand f (x) is bounded, is called an improper integral of the first kind or an infinite integral. Thus, $\int_0^\infty \frac{dx}{1+x^2}$ is an improper integral of the first kind since the upper limit of integration is infinite and the integrand $1/(1 + x^2)$ is bounded. Similarly, $\int_{-\infty}^0 e^x dx$ is an example of an improper integral of the first kind because here the lower limit of integration is infinite. Also $\int_{-\infty}^\infty \frac{dx}{1+x^2}$ is an improper integral of the first kind.

In case the *interval (a, b) is infinite and the integrand f (x) is bounded,* we define

(i) $\int_a^\infty f(x)dx = \lim_{x \to \infty} \int_a^x f(x)dx$, provided that the limit exists finitely *i.e.*, the limit is equal to a definite real number.

(ii) $\int_{-\infty}^b f(x)dx = \lim_{x \to \infty} \int_{-x}^b f(x)dx$, provided that the limit exists finitely.

(iii) $\int_{-\infty}^\infty f(x)dx = \lim_{x_1 \to \infty} \int_{-x_1}^c f(x)dx + \lim_{x_2 \to \infty} \int_c^{x_2} f(x)dx$ provided that both these limits exist finitely.

7. *Improper Integrals of the Second Kind* : A definite integral $\int_a^b f(x)dx$ in which the range of integration is finite but the integrand f (x) is unbounded at one or more points of the interval $a \le x \le b$, is called an improper integral of the second kind.

Thus $\int_0^4 \frac{dx}{(x-2)(x-3)}$ and $\int_0^1 \frac{1}{x_2} dx$ are improper integrals of the second kind.

In the case of the definite integral

$$\int_a^b f(x)dx,$$

if the range of integration (a, b) is finite and the integrand *f (x) is unbounded at one or more points of the given interval,* we define the value of the integral as follows:

(i) If f (x) is unbounded at x = b only *i.e.*, if f (x) $\to \infty$ as x $\to$ b only, then we define

$$\int_a^b f(x)dx = \lim_{\varepsilon \to 0} \int_a^{b-\varepsilon} f(x)dx,$$ provided that the limit exists finitely.

Here ε is a small positive number.

(ii) If f (x) $\to \infty$ as x $\to$ a only, then we define

$$\int_a^b f(x)dx = \lim_{\varepsilon \to 0} \int_{a+\varepsilon}^{b} f(x)dx,$$ provided that the limit exists finitely.

(iii) If f (x) $\to \infty$ as x $\to$ c only, where a < c < b, then we define

$$\int_a^b f(x)dx = \lim_{\varepsilon \to 0} \int_a^{c-\varepsilon} f(x)dx + \lim_{\varepsilon \to 0} \int_{c+\varepsilon}^{b} f(x)dx,$$ provided that both these limits exist finitely.

(iv) If f (x) is unbounded at both the points and b of the interval (a, b) and is bounded at each other point of this interval, we write

$$\int_a^b f(x)dx = \int_a^c f(x)dx + \int_c^b f(x)dx,$$

where a < c < b and the value of the integral exists only if each of the integrals on the right hand side exists.

3.2 Convergence of Improper Integrals

When the limit of an improper integral as defined above, is a definite finite number, we say that the given integral is *convergent* and the value of the integral is equal to the value of that limit. When the limit is ∞ or $-\infty$, the integral is said to be *divergent i.e.*, the value of the integral does not exist.

In case the limit is neither a definite *number* nor ∞ or $-\infty$, the integral is said to be *oscillatory* and in this case also the value of the integral does not exist *i.e.*, the integral is not convergent. We can define the convergence of the infinite integrand $\int_a^\infty f(x)dx$ as follows:

Definition: *The integral* $\int_a^{\infty} f(x)dx$ *is said to converge to the value I, if for any arbitrarily chosen positive number ε, however small but not zero, there exists a corresponding positive number N such that*

$$\left|\int_a^b f(x)dx - I\right| < \varepsilon \text{ for all values of } b \geq N.$$

Similarly, we can define the convergence of an integral, when the lower limit is infinite, or when the integrand becomes infinite at the upper or lower limit.

3.3 Tests for Convergence of Improper Integrals of the First Kind

i.e, to test the convergence of improper integrals in which the range of integration is infinite and the integrand is bounded.

If an integral of the form $\int_a^{\infty} f(x)dx$ or $\int_{-\infty}^{b} f(x)dx$ cannot be actually integrated, its convergence is determined with the help of the following tests:

3.4 Comparison Test

Let f (x) and g (x) be two functions which are bounded and integrable in the interval (a, ∞). Also let g (x) be positive and $|f(x)| \leq g(x)$ *when* $x_0 > a$. *Then, if* $\int_a^{\infty} g(x)dx$ *is convergent,* $\int_a^{\infty} f(x)dx$ *is also convergent.*

Similarly if $|f(x)| \geq g(x)$ *for all values of x greater than some number x0 < a and* $\int_a^{\infty} g(x)dx$ *is divergent, then* $\int_a^{\infty} g(x)dx$ *is also divergent.*

Alternative Form of the Above Comparison Test

If $\lim_{x \to \infty} \frac{f(x)}{g(x)}$ *is a definite number, other than zero, the integrals* $\int_a^{\infty} f(x)dx$ *and* $\int_c^{\infty} g(x)dx$ *either both converge of both diverge.*

Note: While applying comparison test, we generally take $g(x) = \frac{1}{x^n}$

i.e., $\int_a^{\infty} \frac{dx}{x^n}$ is generally taken as the comparison integral.

Theorem:

The comparison integral $\int_a^{\infty} \frac{dx}{x^n}$, *where a > 0, is convergent when > 1 and divergent when n ≤ 1.*

Proof:

By the definition of an improper integral, we have

$$\int_a^{\infty} \frac{dx}{x^n} = \lim_{x \to \infty} \int_a^x \frac{dx}{x^n} = \lim_{x \to \infty} \int_a^x x^{-n} dx$$

$$= \lim_{x \to \infty} \left[\frac{x^{1-n}}{1-n} \right]_a^x, \text{ if } n \neq 1$$

$$= \lim_{x \to \infty} \left[\frac{x^{1-n}}{1-n} - \frac{a^{1-n}}{1-n} \right]. \qquad ...(1)$$

If $n > 1$, then $1 - n$ is negative and so $n - 1$ is positive.

Therefore in this case $\lim_{x \to \infty} x^{1-n} = \lim_{x \to \infty} \frac{1}{x^{n-1}} = \frac{1}{\infty} = 0.$

Hence from (1), we have

$$\int_a^{\infty} \frac{dx}{x^n} = \frac{a^{1-n}}{n-1}, \text{ if } n > 1.$$

Hence the given integral is convergent when $n > 1$.

If $n < 1$, then $1 - n$ is positive

and so $\lim_{x \to \infty} x^{1-n} = \infty$.

$\therefore$ Form (1), we have $\int_a^{\infty} \frac{dx}{x^n} = \infty$.

Hence the given integral is divergent when $n < 1$.

When $n = 1$, we have $\int_a^{\infty} \frac{dx}{x^n} = \int_a^{\infty} \frac{dx}{x}$

$$= \lim_{x \to \infty} \int_a^x \frac{dx}{x} = \lim_{x \to \infty} [\log x]_a^x = \lim_{x \to \infty} [\log x - \log a]$$

$$= \infty - \log a = \infty.$$

Hence the given integral is divergent when $n = 1$.

$\therefore \int_a^{\infty} \frac{dx}{x^n}$ converges when $n > 1$ and diverges when $n \leq 1$.

3.5 The μ-Test

Let $f(x)$ be bounded and integrable in the interval (a, ∞) where $a > 0$.

If there is a number $m > 1$, such that $\lim_{x \to \infty} x^{\mu} f(x)$ exists, then $\int_a^{\infty} f(x)dx$ is convergent.

If there is a number $\mu \leq 1$, such that $\lim_{x \to \infty} x^{\mu} f(x)$ exists and is non-zero, then $\int_a^{\infty} f(x)dx$ is divergent and the same is true if $\lim_{x \to \infty} x^{\mu} f(x)$ is $+\infty$ or $-\infty$.

While applying the μ-test, the value of μ is usually taken to be equal to the highest power of x in the denominator of the integrand minus the highest power of x in the numerator of the integrand.

3.6 Able's Test for the Convergence of Integral of a Product

If $\int_a^\infty f(x)dx$ *converges and* $\phi(x)$ *is bounded and monotonic for* $x > a$, *then* $\int_a^\infty f(x)dx\, f(x)\, dx$ *is convergent.*

3.7 Dirichlet's Test for the Convergence of Integral of a Product

If $f(x)$ *be bounded and monotonic in the interval* $a \le x < \infty$ *and if* $\lim_{x\to\infty} f(x) = 0$, *then the integral* $\int_a^\infty f(x)\phi(x)dx$ *converges provided* $\left|\int_a^x \phi(x)dx\right|$ *is bounded as x takes all finite values.*

6.8 Absolute Convergence

The infinite integral $\int_a^\infty f(x)dx$ *is said to be absolutely convergent if the integral* $\int_a^\infty |f(x)|dx$ *is convergent.*

If the integral $\int_a^\infty f(x)dx$ is absolutely convergent, it is necessarily convergent. But if the integral $\int_a^\infty f(x)dx$ is convergent, it is not necessarily absolutely convergent. Thus, absolute convergence gives a sufficient but not a necessary condition for the convergence of an infinite integral.

3.9 Tests for Convergence of Improper Integrals of the Second Kind

Now we shall make a study of the tests for the convergence of a definite of the type $\int_a^b f(x)dx$ in which the range of integration is finite and the integrand $f(x)$ is unbounded at one or more points of the given interval [a, b]. It is sufficient to consider the case when $f(x)$ becomes abounded at $x = a$ and bounded for all other values of x in the interval [a, b]. In this case we have

$$\int_a^b f(x)dx = \lim_{\varepsilon\to 0}\int_{a+\varepsilon}^b f(x)dx.$$

In the articles to follow we give a few important tests for the convergence of the above integral.

3.10 Comparison Test

Consider the improper integral $\int_a^b f(x)dx$, *where the range of integration* (*a, b*) *is finite and f (x) is unbounded only at x = a. Let g (x) be positive in the interval* (*a +* ε, *b*) *and* $|f(x)| \leq g(x)$ *in the interval* (*a +* ε, *b*). *Then* $\int_a^b f(x)dx$ *is convergent if* $\int_a^b g(x)dx$ *is convergent.*

Similarly if $|f(x)| \geq g(x)$ *for all values of x in the interval* (*a + e, b*), *then* $\int_a^b f(x)dx$ *is divergent provided* $\int_a^b g(x)dx$ *is divergent.*

Alternative Form of the Above Comparison Test

If $\lim_{x \to a} \frac{f(x)}{g(x)}$ is a definite number, other than zero, the integrals $\int_a^b f(x)dx$ and $\int_a^b g(x)dx$ either both converge or both diverge.

Note: While applying the above comparison test, we generally take $g(x) = \frac{1}{(x-a)^n}$ *i.e.*, $\int_a^b \frac{dx}{(x-a)^n}$ is generally taken as the comparison integral.

Theorem:

The comparison integral $\int_a^b \frac{dx}{(x-a)^n}$ *is convergent when n < 1 and divergent when n ≥ 1.*

Proof:

We have

$$\int_a^b \frac{dx}{(x-a)^n} = \lim_{\varepsilon \to 0} \int_{a+\varepsilon}^b \frac{dx}{(x-a)^n}$$

$$= \lim_{\varepsilon \to 0} \int_{a+\varepsilon}^b (x-a)^{-n} dx$$

$$= \lim_{\varepsilon \to 0} \left[\frac{(x-a)^{-n+1}}{1-n} \right]_{a+\varepsilon}^b \quad \text{if } n \neq 1$$

$$= \lim_{\varepsilon \to 0} \left[\frac{(b-a)^{1-n}}{1-n} - \frac{\varepsilon^{1-n}}{1-n} \right]. \qquad ...(1)$$

If n < 1, then 1 – n is positive and so $\lim_{\varepsilon \to 0} \varepsilon^{1-n} = 0$. Therefore from (1), we have

$$\int_a^b \frac{dx}{(x-a)^n} = \frac{(b-a)^{1-n}}{1-n}, \textit{ if } n < 1.$$

Hence the given integral converges when $n < 1$.

If $n > 1$, then $1 - n$ is negative and so $n - 1$ is positive. Therefore in this case, from (1), we have

$$\int_a^b \frac{dx}{(x-a)^n} = \lim_{\varepsilon \to 0}\left[\frac{(b-a)^{1-n}}{1-n} + \frac{1}{(n-1)\varepsilon^{n-1}}\right] = \infty.$$

Hence the given integral diverges when $n > 1$.

when $n = 1$, we have

$$\int_a^b \frac{dx}{(x-a)^n} = \int_a^b \frac{dx}{(x-a)}$$

$$= \lim_{\varepsilon \to 0}\int_{a+\varepsilon}^b \frac{dx}{x-a} = \left[\log(x-a)\right]_{a+\varepsilon}^b$$

$$= \lim_{\varepsilon \to 0} [\log(b-a) - \log \varepsilon]$$

$$= \infty, \qquad [\because \log 0 = -\infty]$$

Hence, the given integral diverges when $x = 1$.

3.11 The μ-Test

Let $f(x)$ be unbounded at $x = a$ and be bounded and integrable in the arbitrary interval

$(a + \varepsilon, b)$, where $0 < \varepsilon < b - a$.

If there is a number μ between 0 and 1 such that

$\lim_{x \to a+0} (x - a)^\mu f(x)$ *exists, then* $\int_a^b f(x)dx$ *is convergent.*

If there is a number $\mu \geq 1$ such that $\lim_{x \to a+0} (x - a)^\mu f(x)$ *exist and is non-zero, then* $\int_a^b f(x)dx$ *is divergent and the same is true if* $\lim_{x \to a+0} (x - a)^\mu f(x) = +\infty$ *or* $-\infty$.

In case $f(x)$ is unbounded at $x = b$, we should find

$$\lim_{x \to b+0} (b - x)^\mu . f(x),$$

the other conditions of the test remaining the same.

3.12 Abel's Test

If $\int_a^b f(x)dx$ *converges and* $\phi(x)$ *is bounded and monotonic for* $a \le x \le b$, *then* $\int_a^b f(x)\phi(x)dx$ *converges.*

3.13 Dirichlet's Test

If $\int_{a+\varepsilon}^b f(x)dx$ *be bounded and* $\phi(x)$ *be bounded and monotonic on the interval* $a \le x \le b$, *converging to zero as x tends to a, then* $\int_a^b f(x)\phi(x)dx$ *converges.*

MISCELLANEOUS EXAMPLES

Example 1:

Test the convergence of the integral $\int_1^\infty \frac{dx}{\sqrt{(x^3+1)}}dx$.

Solution:

Let $f(x) = \frac{1}{\sqrt{(x^3+1)}} = \frac{1}{x^{3/2}\sqrt{\{1+(1/x^2)\}}}$.

Take $g(x) = \frac{1}{x^{3/2}}$. We have $\lim_{x\to\infty}\frac{f(x)}{g(x)} = \lim_{x\to\infty}\frac{1}{\sqrt{\{1+(1/x^3)}} = 1$,

which is finite and non-zero. Therefore $\int_1^\infty f(x)dx$ and $\int_1^\infty g(x)dx$ are either both convergent or both divergent. But the comparison integral $\int_1^\infty g(x)dx$ *i.e.*, $\int_1^\infty \frac{dx}{x^{3/2}}$ is convergent because here n = 3/2 which is > 1.

Hence $\int_1^\infty f(x)dx$ *i.e.*, $\int_1^\infty \frac{dx}{\sqrt{(x^3+1)}}$ is also convergent.

Example 2:

Test the convergence of

(i) $\int_2^\infty \frac{dx}{\sqrt{(x^2-x-1)}}$;

(ii) $\int_2^\infty \frac{dx}{\sqrt{(x^2-1)}}$.

Solution:

(i) Let $f(x) = \dfrac{1}{\sqrt{(x^2 - x - 1)}} = \dfrac{1}{\sqrt{\left\{\left(x - \frac{1}{2}\right)^2 - \frac{5}{4}\right\}}}$

Obviously f (x) = is bounded in the interval $(2, \infty)$.

We can writ $f(x) = \dfrac{1}{x\sqrt{\{1 - (1/x) - (1/x^2)\}}}$ Take $g(x) = \dfrac{1}{x}$.

We have

$$\lim_{x \to \infty} \frac{f(x)}{g(x)} = \lim_{x \to \infty} \frac{1}{\sqrt{\{1 - (1/x) - (1/x^2)\}}} = 1,$$

which is finite and non-zero. Therefore

$$\int_2^\infty f(x)dx \text{ and } \int_2^\infty g(x)dx$$

either bot converge or both diverge.

But the comparison integral

$$\int_2^\infty g(x)dx \text{ } i.e., \int_2^\infty \frac{1}{x}dx$$

is divergent because here n = 1. Hence

$$\int_2^\infty f(x)dx \, i.e., \int_2^\infty \frac{dx}{\sqrt{(x^2 - x - 1)}}$$

is also divergent.

(ii) Proceed exactly as inpart (i).

Here also the given integral is divergent.

Example 3:

Test the convergence of $\displaystyle\int_0^\infty \frac{x^3}{(x^2 + a^2)^2}dx$.

Solution:

Let $f(x) = \dfrac{x^3}{(x^2 + a^2)^2}dx$. Then f (x) is bounded in the interval $(0, \infty)$.

Let b > 0.

We can write

$$\int_0^\infty \frac{d^2}{(x^2 + a^2)^2}dx = \int_0^b \frac{x^3}{(x^2 + a^2)^2}dx + \int_b^\infty \frac{x^3}{(x^2 + a^2)}dx.$$

The first integral on the right hand side is a proper integral because the interval of integration (0, b) is finite and the integrand f (x) is bounded in this interval. So we need to check the convergence of the integral

$$\int_b^{\infty} \frac{x^3}{(x^2+a^2)^2}\,dx \text{ only.}$$

We can write

$$f(x) = \frac{x^3}{x^4\{1+(a^2/x^2)\}^2} = \frac{1}{x\{1+(a^2/x^2)\}^2}.$$

Take g (x) = 1/x.

We have $\lim_{x\to\infty} \frac{f(x)}{g(x)} = \lim_{x\to\infty} \frac{1}{\{1+(a^2/x^2)\}^2} = 1$, which is finite and non-zero. Therefore $\int_b^{\infty} f(x)dx$ and $\int_b^{\infty} g(x)dx$ either both converge or both diverge. But the comparison integral $\int_b^{\infty} g(x)dx$ *i.e.*, $\int_b^{\infty} \frac{1}{x}dx$ is divergent. Therefore, $\int_b^{\infty} f(x)dx$ is also divergent. Hence $\int_0^{\infty} \frac{x^3}{(x^2+a^2)^2}dx$ is divergent.

Example 4:

Test the convergence of

$$\int_0^{\infty} e^{-x}\frac{\sin x}{x}\,dx.$$

Solution:

We can write

$$\int_0^{\infty} e^{-x}\frac{\sin x}{x}dx = \int_0^{1} e^{-x}\frac{\sin x}{x}dx + \int_1^{\infty} e^{-x}\frac{\sin x}{x}dx.$$

Since $\lim_{x\to\infty} e^{-x}\frac{\sin x}{x} = 1$, therefore the integrand $e^{-x}\frac{\sin x}{x}$ is bounded throughout the finite interval (0, 1). So $\int_0^1 e^{-x}\frac{\sin x}{x}$ is a proper integral and therefore it is convergent. Thus we need to check the convergence of $\int_1^{\infty} e^{-x}\frac{\sin x}{x}dx$ only.

Let $f(x) = e^{-x}\frac{\sin x}{x}$. Then f (x) is bounded in the interval (1, ∞).
Take g (x) = e^{-x}. Then g (x) is positive in the interval (1, ∞).

We have

$$|f(x)| = \left|e^{-x}\frac{\sin x}{x}\right| = e^{-x}.\ |\sin x|.\ \frac{1}{2}$$

$$\le e^{-x}, \text{ since } |\sin x| \le 1 \text{ and } \frac{1}{x} \le 1.$$

Thus $|f(x)|$ $g(x)$ throughout the interval $(1, \infty)$.

∞ by comparison test $\int_1^\infty f(x)dx$ is convergent if $\int_1^\infty g(x)dx$ is convergent.

Now $\int_1^\infty g(x)dx = \int_1^\infty e^{-x}dx = \lim_{x\to\infty}\int_1^x e^{-x}dx$

$$= \lim_{x\to\infty}\left[-e^{-x}\right]_1^x$$

$$= \lim_{x\to\infty}\ [-e^{-x} + e^{-1}] = 0 + e^{-1} = 1/e,$$

which is a definite finite number. Hence $\int_1^\infty g(x)dx$ is convergent.

$\therefore\ \int_1^\infty f(x)dx$ is also convergent.

Hence $\int_0^\infty e^{-x}\frac{\sin x}{x}dx$ is convergent because the sum of two convergent integrals is also convergent.

Example 5:

Test the convergence of the integral

$$\int_0^\infty \frac{1-\cos x}{x^2}dx.$$

Solution:

We have

$$\int_0^\infty \frac{1-\cos x}{x^2}dx = \int_0^\infty \frac{2\sin^2(x/2)}{x^2}dx$$

$$= \int_0^\infty \frac{2\sin^2 t}{4t^2}\ .\ 2dt,$$

putting $\frac{x}{2} = t$

so that $dx = 2dt$

$$= \int_0^\infty \frac{\sin^2 t}{t^2}dt.$$

Example 6:

Show that the integral $\int_0^\infty e^{-x^2}dx$ *is convergent.*

Solution:

We have

$$\int_0^\infty e^{-x^2}dx = \int_0^1 e^{-x^2}dx + \int_1^\infty e^{-x^2}dx.$$

Obviously $\int_0^1 e^{-x^2}dx$ is a proper integral because here the interval of integration (0, 1) is finite and the integrand e^{x^2} is bounded throughout this interval. Therefore this integral is convergent. So we need to check the convergence of $\int_1^\infty e^{-x^2}dx$ only.

Let $f(x) = e^{-x^2}$. Take $g(x) = xe^{-x^2}$ so that g (x) is positive throughout the interval $(1, \infty)$. We have

$$|f(x)| = e^{-x^2} \le xe^{-x^2}, \text{ since } x \ge 1.$$

Thus $|f(x)| \le g(x)$ throughout the interval $(1, \infty)$.

$\therefore$ by comparison test $\int_1^\infty e^{-x^2}dx$ is convergent if $\int_1^\infty xe^{-x^2}dx$ is convergent.

$$\text{Now } \int_1^\infty xe^{-x^2}dx = \lim_{X\to\infty}\int_1^X xe^{-x^2}dx$$

$$= \lim_{X\to\infty}\left(-\frac{1}{2}e^{-x^2}\right)_1^X = \lim_{X\to\infty}\left(-\frac{1}{2}e^{-X^2} + \frac{1}{2}e^{-1}\right)$$

$$= \frac{1}{2}e^{-1}, \text{ which is a definite number.}$$

$\therefore \int_1^\infty xe^{-x^2}dx$ is convergent and so $\int_1^\infty e^{-x^2}dx$ is also convergent

Hence the given integral $\int_0^\infty e^{-x^2}dx$ is also convergent as it is the sum of two convergent integrals.

Example 7:

Examine the convergence of $\int_1^\infty \frac{dx}{x^{1/3}(1+x^{1/2})}$.

Solution:

$$\text{Let } f(x) = \int_1^\infty \frac{dx}{x^{1/3}(1+x^{1/2})} = \frac{1}{x^{1/3}x^{1/2}\{1+(1/x^{1/2})\}}$$

$$= \frac{1}{x^{5/6}\{1+(1/x^{1/2})\}}.$$

Obviously $f(x)$ is bounded in the integral $(1, \infty)$.

Take $\mu = \frac{5}{6} - 0 = \frac{5}{6}$. We have

$$\lim_{x \to \infty} x^{\mu} f(x) = \lim_{x \to \infty} x^{5/6} \cdot \frac{1}{x^{5/6}\{1+(1/x^{1/2})\}}$$

$$= \lim_{x \to \infty} \frac{1}{1+(1/x^{1/2})} = 1,$$

which is finite and non-zero. Since $\mu = \frac{5}{6}$ *i.e.*, < 1, it follows from the μ-test that the given integral is divergent.

Example 8:

Examine the convergence of $\int_0^\infty \frac{x\,dx}{(1+x)^3}$.

Solution:

Let $a > 0$. Then we have

$$\int_0^\infty \frac{x\,dx}{(1+x)^3} = \int_0^a \frac{x\,dx}{(1+x)^3} + \int_a^\infty \frac{x\,dx}{(1+x)^3}.$$

The first integral on the right hand side is convergent because it is a proper integral. We observe that in this in this integral the range of integrating $(0, a)$ is finite and the integrand $x/(1+x)^3$ is bounded throughout the interval $(0, a)$. So we need to check the convergence of $\int_a^\infty \frac{x\,dx}{(1+x)^3}$ only.

Let $f(x) = \frac{x}{(1+x)^3}$. Then $f(x)$ is bounded in the interval (a, ∞).Take $\mu = 3 - 1 = 2$. Then

$$\lim_{x \to \infty} x^{\mu} f(x) = \lim_{x \to \infty} x^2 \cdot \frac{x}{(1+x)^3} = \lim_{x \to \infty} \frac{1}{\{1+(1/x)\}^3} = 1,$$

which exists *i.e.*, is equal to a definite real number.

Since $\mu = 2$ *i.e.*, > 1,

therefore by μ-test the integral $\int_a^\infty \frac{x\,dx}{(1+x)^3}$ is convergent.

Hence $\int_0^\infty \frac{x\,dx}{(1+x)^3}$ is also convergent because it is the sum of two convergent integrals.

Example 9:

Show that the following integrals are convergent.

$$\text{(i)} \int_0^\infty \frac{x^2}{(a^2+x^2)^2} dx; \quad \text{(ii)} \int_1^\infty \frac{dx}{(1+x)\sqrt{x}}.$$

Solution:

(i) Let b > 0. Then we have

$$\int_0^\infty \frac{x^2}{(a^2+x^2)^2} dx = \int_0^b \frac{x^2}{(a^2+x^2)^2} dx + \int_b^\infty \frac{x^2}{(a^2+x^2)^2} dx.$$

The first integral on the right hand side is convergent because it is a proper integral. So we need to check the convergence of

$$\int_b^\infty \frac{x^2}{(a^2+x^2)^2} dx.$$

Let $f(x) = \dfrac{x^2}{(a^2+x^2)^2}$. Then f (x) is bounded in the integrals (b, ∞).

Take $\mu = 4 - 2 = 2$. Then $\lim_{x \to \infty} x^\mu f(x) = \lim_{x \to \infty} x^2 \cdot \dfrac{x^2}{(a^2+x^2)^2} =$

$\lim_{x \to \infty} \dfrac{1}{\{1+(a^2/x^2)\}^2} = 1,$

which is a definite real number.

Since $\mu > 1$, therefore by μ-test $\displaystyle\int_b^\infty \frac{x^2}{(a^2+x^2)^2} dx$ is convergent.

Hence $\displaystyle\int_0^\infty \frac{x^2 dx}{(a^2+x^2)^2}$ is also convergent because it is the sum of two convergent integrals.

(ii) Apply μ-test by taking $\mu = \dfrac{3}{2}$.

Example 10:

Test the convergence of $\displaystyle\int_b^\infty \frac{x^{3/2}}{\sqrt{(x^4 - a^4)}}$, *where b > a.*

Solution:

Let $f(x) = \dfrac{x^{3/2}}{\sqrt{(x^4 - a^4)}}$. Then f (x) is bounded in the interval (b. ∞).

Take $\mu = 2 - \dfrac{3}{2} = \dfrac{1}{2}$. Then $\lim_{x \to \infty} x^\mu f(x) = \lim_{x \to \infty} x^{1/2} \cdot \dfrac{x^{3/2}}{x^2 \sqrt{\{1-(a^4 x^4)\}}}$

$= \lim_{x \to \infty} \dfrac{1}{\sqrt{\{1-(a^4/x^4)\}}} = 1,$ which is finite and non-zero.

Since $\mu < 1$, therefore by μ-test the given integral is divergent.

Example 11:

Examine the convergence of $\int_a^\infty \frac{dx}{x(\log x)^{n+1}}$, *where* $a > 1$.

Solution:

Let $\log x = t$ so that $(1/x)\,dx = dt$.

$$\therefore \int_a^\infty \frac{dx}{x(\log x)^{n+1}} = \int_{\log a}^\infty \frac{dt}{t^{n+1}}.$$

Let $f(t) = 1/t^{n+1}$. Then $f(t)$ is bounded in the interval $(\log a, \infty)$.

Take $\mu = (n + 1) - 0 = n + 1$. Then

$$\lim_{t \to \infty} t^\mu f(t) = \lim_{t \to \infty} \frac{t^{n+1}}{t^{n+1}} = \lim_{t \to \infty} 1 = 1,$$

which is finite and non-zero.

Therefore by μ-test, the given integral is convergent if

$\mu > 1$ *i.e.*, $n + 1 > 1$ *i.e.*, $n > 0$

and divergent if $\mu \le 1$ *i.e.*, $n + 1 \le 1$ *i.e.*, $n \le 0$.

Example 12:

Show that the integral $\int_1^\infty x^{n-1}e^{-x}dx$ *is convergent.*

Solution:

Let $f(x) = x^{n-1} e^{-x}$. Then $f(x)$ is bounded in the interval $(1, \infty)$. We have

$$\lim_{x \to \infty} x^\mu f(x) = \lim_{x \to \infty} \frac{x^\mu . x^{n-1}}{e^x} = \lim_{x \to \infty} \frac{x^{\mu+n-1}}{1 + x + \frac{x^2}{2!} + \ldots} = 0 \text{ for all values}$$

of μ and n.

Taking $\mu > 1$, we see by μ-test that the integral $\int_1^\infty x^{n-1}e^{-x}dx$

is convergent for all values of n.

Example 13:

Test the convergence of

$$\int_a^\infty (1 - e^{-x}) \frac{\cos x}{x^2} dx, \text{ when } a > 0.$$

Solution:

Let $f(x) = \frac{\cos x}{x^2}$ and $\phi(x) = 1 - e^{-x}$.

We have $\left|\dfrac{\cos x}{x^2}\right| \leq \dfrac{1}{x^2}$ as $|\cos x| \leq 1$.

Since $\int_a^\infty \dfrac{1}{x^2}dx$ is convergent, therefore by comparison test $\int_a^\infty \dfrac{\cos x}{x^2}dx$ is also convergent.

Again $\phi(x) = 1 - e^{-x}$ is monotonic increasing and bounded function for $x > a$.

Hence by Able's test $\int_a^\infty (1-e^{-x})\dfrac{\cos x}{x^2}dx$ is convergent.

Example 14:

Test the convergence of $\int_a^\infty e^{-x}\dfrac{\sin x}{x^2}dx$ *where* $a > 0$.

Solution:

Let $f(x) = \dfrac{\sin x}{x^2}$ and $\phi(x) = e^{-x}$.

Since $\left|\dfrac{\sin x}{x^2}\right| \leq \dfrac{1}{x^2}$ and $\int_a^\infty \dfrac{1}{x^2}dx$ is convergent.

Again e^{-x} is monotonic decreasing and bounded function for $x > a$.

Hence by Able's test $\int_a^\infty e^{-x}\dfrac{\sin x}{x^2}dx$ is convergent, therefore by comparison test $\int_a^\infty \dfrac{\sin x}{x^2}dx$ is also convergent.

Example 15:

Test the convergence of the integral

$$\int_a^\infty \frac{\sin x}{\sqrt{x}}dx, \text{ where } a > 0.$$

Solution:

Let $f(x) = \dfrac{1}{\sqrt{x}}$ and $\phi(x) = \sin x$.

Now $\dfrac{1}{\sqrt{x}}$ is bounded and monotonic decreasing for all $x \geq a$ and $\lim_{x \to \infty} \dfrac{1}{\sqrt{x}} = 0$.

Also $\left|\int_a^x \phi(x)dx\right| = \left|\int_a^x \sin x\, dr\right| = |\cos a - \cos x| \leq 2$, for all finite values of x. [Note that the value of cos x lies between – 1 and 1].

$\therefore \left|\int_a^X \phi(x)dx\right|$ is bounded for all finite values of x.

Hence by Dirichlet's test the integral $\int_0^\infty \frac{\sin x}{\sqrt{x}} dx$ is convergent.

Example 16:

Show that the integral $\int_a^\infty x^{n-1}e^{-x}dx$ *is convergent, where a > 0.*

Solution:

Do your self.

Example 17:

Test the convergence of the integral

$$\int_0^\infty \frac{x^{2m}}{1+x^{2n}} dx,$$

where m and n are positive integers.

Solution:

Let a > 0. We have

$$\int_0^\infty \frac{x^{2m}}{1+x^{2n}} dx = \int_0^a \frac{x^{2x}}{1+x^{2n}} dx + \int_a^\infty \frac{x^{2m}}{1+x^{2n}} dx.$$

The first integral on the right hand side is a proper integral and so it is convergent. Therefore the given integral is convergent or divergent according as $\int_a^\infty \frac{x^{2m}}{1+x^{2n}} dx$ is convergent or divergent.

To test the convergence of $\int_a^\infty \frac{x^{2m}}{1+x^{2n}} dx$,

let us take m = 2n – 2m.

We have $\lim_{x\to\infty} x^{\mu} \frac{x^{2m}}{1+x^{2n}} = \lim_{x\to\infty} x^{2n-2m} \frac{x^{2m}}{x^{2n}\{1+(1/x^{2n})\}}$

$$= \lim_{x\to\infty} \frac{1}{1+(1/x^{2n})} = 1, \text{ which is finite and non-zero.}$$

$\therefore$ by μ-test, the given integral is convergent if μ > 1 *i.e.*, if 2n – 2m > 1 which is possible if n > m since m and n are positive integers. Also by μ-test, the given integral is divergent if μ ≤ 1 *i.e.*, if 2n – 2m ≤ 1 *i.e.*, if n ≤ m since n and m are positive integers.

Example 18:

Show that $\int_a^\infty \sin x^2 dx$ *is convergent.*

Solution:

We have $\int_0^\infty \sin x^2 dx = \int_0^1 \sin x^2 dx + \int_1^\infty \sin x^2 dx$.

But $\int_0^1 \sin x^2 dx$ is a proper integral and hence convergent.

Now it remains to test the convergence of $\int_1^\infty \sin x^2 dx$. We can write

$$\int_1^\infty \sin x^2 dx = \int_1^\infty 2x.(\sin x^2).\frac{1}{2x}dx.$$

Let $f(x) = \dfrac{1}{2x}$ and $\phi(x) = 2x \sin x^2$.

The function $f(x) = \dfrac{1}{2x}$ is bounded and monotonic decreasing for all $x \geq 1$ and $\lim_{x \to \infty} \dfrac{1}{2x} = 0$.

Also $\left|\int_1^X \phi(x)dx\right| = \int_1^X 2x \sin x^2 dx$

$= |\cos 1^2 - \cos x^2| \leq 2$, for all finite values of x.

$\therefore \left|\int_1^X \phi(x)dx\right|$ is bounded for all finite values of x.

Hence by Dirichlet's test

$$\int_1^\infty \frac{1}{2x}.(\sin x^2)2x\,dx \quad i.e., \int_1^\infty \sin x^2 dx \text{ is convergent.}$$

Since the sum of two convergent. Integrals is convergent, therefore the integral $\int_0^\infty \sin x^2 dx$ is convergent.

Example 19:

Show that the integral $\int_0^\infty \frac{\sin x}{x} dx$ *is convergent.*

Solution:

We have $\int_0^\infty \frac{\sin x}{x} dx = \int_0^a \frac{\sin x}{x} dx + \int_a^\infty \frac{\sin x}{x} dx$, where $a > 0$.

Since $\lim_{x \to \infty} \dfrac{\sin x}{x} = 1$, the integral $\int_0^a \frac{\sin x}{x} dx$ is a proper integral and hence convergent.

Now to test the convergence of $\int_a^\infty \frac{\sin x}{x} dx$.

Let $f(x) = 1/x$ and $\phi(x) = \sin x$.

The function $f(x) = 1/x$ is bounded and monotonic decreasing for all $x \geq a$ and $\lim_{x \to \infty} \frac{1}{x} = 0$.

Also $\left|\int_a^x \phi(x)\, dx\right| = \left|\int_a^x \sin x\, dx\right| = |\cos a - \cos x| \leq 2$, for all finite values of x.

$\therefore \left|\int_a^x \phi(x)\, dx\right|$ is bounded for all finite values of x.

Hence by Dirichlet's test the integral $\int_a^\infty \frac{\sin x}{x} dx$ is convergent.

Since the sum of two convergent integrals is convergent, therefore $\int_0^\infty \frac{\sin x}{x} dx$ is convergent.

Example 20:

Prove that $\int_a^\infty \frac{\cos \alpha x - \cos \beta x}{x} dx$ *is convergent where a > 0.*

Solution:

We have

$$\int_a^\infty \frac{\cos \alpha x - \cos \beta x}{x} dx = \int_a^\infty \frac{\cos \alpha x}{x} dx - \int_a^\infty \frac{\cos \beta x}{x} dx.$$

The fiction $f(x) = 1/x$ is bounded and monotonic decreasing for all x $\geq a$ and $\lim_{x \to \infty} \frac{1}{x} = 0$.

Also $\left|\int_a^x \cos \alpha x\, dx\right| = \left|\frac{1}{\alpha}(\sin \alpha x - \sin \alpha a\right| \leq \frac{2}{|\alpha|}$.

$\therefore \left|\int_a^x \cos \beta x\, dx\right|$ is bounded for all finite values of x.

Similarly $\left|\int_a^x \cos \beta x\, dx\right|$ is bounded for all finite values of x.

$\therefore$ by Dirichlet's test both the integrals

$\int_a^\infty \frac{\cos \alpha x}{x} dx$ and $\int_a^\infty \frac{\cos \beta x}{x} dx$ are convergent.

Hence the given integral is convergent.

Example 21:

Show that the integral

$$\int_a^{\infty} e^{-\alpha x} \frac{\sin x}{x} dx, \; a \geq 0 \text{ is convergent.}$$

Solution:

We have

$$\int_a^{\infty} e^{-\alpha x} \frac{\sin x}{x} dx = \int_0^{\alpha} e^{-\alpha x} \frac{\sin x}{x} dx + \int_{\alpha}^{\infty} e^{-\alpha x} \frac{\sin x}{x} dx,$$

where $\alpha > 0$.

Since $\lim_{x \to 0} e^{-\alpha x} \frac{\sin x}{x} = 1$, the integral $\int_0^{\alpha} e^{-\alpha x} \frac{\sin x}{x} dx$ is a proper integral and hence convergent.

Now it remains to test the convergence of

$$\int_{\alpha}^{\infty} e^{-\alpha x} \frac{\sin x}{x} dx.$$

Let $f(x) = \frac{e^{-ax}}{x}$ and $f(x) = \sin x$.

Obviously the function $f(x) = \frac{1}{xe^{ax}}$ is bounded and monotonic decreasing

for all $x \geq \alpha$ and $\lim_{x \to \infty} f(x) = \lim_{x \to \infty} \frac{1}{xe^{ax}} = 0$.

Moreover $\left|\int_{\alpha}^{x} \phi(x)dx\right| = \left|\int_{\alpha}^{x} \sin x\, dx\right| = |\cos a - \cos x| \leq 2$, for all finite values of x.

$\therefore \left|\int_{\alpha}^{x} \phi(x)dx\right|$ is bounded for all finite values of x.

$\therefore$ by Dirichlet's test $\int_{\alpha}^{\infty} e^{-ax} \frac{\sin x}{x} dx$ is convergent.

Since the sum of two convergent integrals is convergent, therefore $\int_0^{\infty} e^{-ax} \frac{\sin x}{x} dx$ is convergent.

Example 22:

Show that $\int_1^{\infty} \frac{\sin x}{x^4} dx$ *is absolutely convergent.*

Solution:

The integral $\int_1^{\infty} \frac{\sin x}{x^4} dx$ will be absolutely convergent if $\int_1^{\infty} \left|\frac{\sin x}{x^4}\right| dx$ is convergent.

Let $f(x) = \left|\frac{\sin x}{x^4}\right|$. Then $f(x)$ is bounded in the interval $(1, \infty)$. We have

$$f(x) = \left|\frac{\sin x}{x^4}\right| = \frac{|\sin x|}{x^4} \le \frac{1}{x^4}, \text{ since } |\sin x| \le 1.$$

$\therefore$ by comparison test, $\int_1^\infty f(x)dx$ is convergent if $\int_1^\infty \frac{1}{x^4}dx$ is convergent.

But the comparison integral $\int_1^\infty \frac{1}{x^4}dx$ is convergent because here $n = 4$ which is > 1.

Hence $\int_1^\infty f(x)dx$ is convergent and so the given integral is absolutely convergent.

Example 23:

Show that $\int_0^\infty \frac{\sin mx}{a^2 + x^2}dx$ *converges absolutely.*

Solution:

The integral $\int_0^\infty \frac{\sin mx}{a^2 + x^2}dx$ will be absolutely convergent if $\int_0^\infty \left|\frac{\sin mx}{a^2 + x^2}\right|dx$ is convergent.

Let $f(x) = \left|\frac{\sin mx}{a^2 + x^2}\right|$. Then $f(x)$ is bounded in the interval $(0, \infty)$. We have

$$f(x) = \left|\frac{\sin mx}{a^2 + x^2}\right| = \frac{|\sin mx|}{a^2 + x^2} \le \frac{1}{a^2 + x^2}, \text{ since } |\sin mx| \le 1.$$

$\therefore$ by comparison test, $\int_0^\infty f(x)dx$ is convergent if $\int_0^\infty \frac{1}{a^2 + x^2}dx$ is convergent.

$$\text{But } \int_0^\infty \frac{dx}{a^2 + x^2} = \lim_{X \to \infty} \int_0^X \frac{dx}{a^2 + x^2} = \lim_{X \to \infty} \left[\frac{1}{a}\tan^{-1}\frac{x}{a}\right]_0^X$$

$$= \lim_{X \to \infty} \left[\frac{1}{a}\tan^{-1}\frac{X}{a} - 0\right] = \frac{1}{a}\cdot\frac{\pi}{2},$$

which is a definite real number.

$\therefore \int_0^\infty \frac{dx}{a^2 + x^2}$ is convergent. Hence $\int_0^\infty f(x)dx$ is also convergent and so the given integral is absolutely convergent.

Example 24:

Show that the integral $\int_0^{\infty} e^{-x}\cos mx\,dx$ *converges absolutely.*

Solution:

The integral $\int_0^{\infty} e^{-x}\cos mx\,dx$ will be absolutely convergent if $\int_0^{\infty} \left|e^{-x}\cos mx\right|dx$ is convergent.

Let $f(x) = |e^{-x}\cos mx|$. Then $f(x)$ is bounded in the interval $(0, \infty)$. We have

$$f(x) = |e^{-x}\cos mx| = e^{-x}|\cos mx| \le e^{-x}, \text{ since } |\cos mx| \le 1.$$

$\therefore$ by comparison test, $\int_0^{\infty} f(x)dx$ is convergent if $\int_0^{\infty} e^{-x}dx$ is convergent.

But $\int_0^{\infty} e^{-x}dx = \lim_{X\to\infty}\int_0^{X} e^{-x}dx = \lim_{X\to\infty}\left[-e^{-x}\right]_0^{\infty}$

$= \lim_{X\to\infty}[-e^{-X}+1] = 1$, which is a definite real number.

$\therefore \int_0^{\infty} e^{-x}dx$ is convergent.

Hence $\int_0^{\infty} f(x)dx$ is convergent and so the given integral is absolutely convergent.

Example 25:

Discuss the convergence of the following integrals by evaluating them

(i) $\int_1^{\infty}\frac{dx}{\sqrt{x}}$; (ii) $\int_1^{\infty}\frac{dx}{x^{3/2}}$.

Solution:

(i) We have

$$\int_1^{\infty}\frac{dx}{\sqrt{x}} = \lim_{X\to\infty}\int_1^{X}\frac{dx}{\sqrt{x}}, \text{ (By def.)}$$

$$= \lim_{X\to\infty}\int_1^{X} x^{-1/2}dx = \lim_{X\to\infty}\left[\frac{x^{1/2}}{1/2}\right]_1^{X}$$

$$= \lim_{X\to\infty}[2\sqrt{X}-2] = \infty$$

Thus, the limit does not exist finitely and therefore the given integral is divergent (*i.e.*, the integral does not exist).

(ii) We have

$$\int_0^{\infty}\frac{dx}{x^{3/2}} = \lim_{X\to\infty}\int_1^{X}\frac{dx}{x^{3/2}}, \qquad \text{(By def.)}$$

$$= \lim_{X \to \infty} \int_1^X x^{-3/2} dx = \lim_{X \to \infty} \left[\frac{x^{-1/2}}{-1/2} \right]_1^X$$

$$= \lim_{X \to \infty} \left[-\frac{2}{\sqrt{x}} \right]_1^X$$

$$= \lim_{X \to \infty} \left[-\frac{2}{\sqrt{X}} + 2 \right] = 2.$$

Thus, the limit exists and is unique and finite; therefore the given integral is convergent and its value is 2.

Example 26:

Evaluate $\int_1^\infty \frac{dx}{x}$.

Solution:

We have

$$\int_1^\infty \frac{dx}{x} = \lim_{X \to \infty} \int_1^X \frac{dx}{x} = \lim_{X \to \infty} [\log x]_1^X$$

$$= \lim_{X \to \infty} [\log X - 0] = \infty.$$

Thus, the limit does not exist finitely and therefore the given integral is divergent (*i.e.*, the integral does not exist).

Example 27:

Evaluate $\int_3^\infty \frac{dz}{(x-2)^2}$.

Solution:

We have

$$\int_3^\infty \frac{dz}{(x-2)^2} = \lim_{X \to \infty} \int_3^X \frac{dx}{(x-2)^2}, \qquad \text{(By def.)}$$

$$= \lim_{X \to \infty} \int_3^X (x-2)^{-2} dx = \lim_{X \to \infty} \left[\frac{(x-2)^{-1}}{-1} \right]_3^X$$

$$= \lim_{X \to \infty} \left[-\frac{1}{x-2} \right]_3^X = \lim_{X \to \infty} \left[-\frac{1}{X-2} + 1 \right] = 1,$$

Which is a definite real number. Therefore the given integral is convergent and its value is 1.

Example 28:

Test the convergence of $\int_0^\infty e^{-mx} dx$, $(m > 0)$.

Solution:

We have $\int_0^\infty e^{-mx}dx = \lim_{x\to\infty}\int_0^x e^{-mx}dx,$ (By def.)

$$= \lim_{x\to\infty}\left[\frac{e^{-mx}}{-m}\right]_0^x = \lim_{x\to\infty}\left\{-\frac{1}{m}(e^{-mx}-1)\right\} = -\frac{1}{m}[0-1] = \frac{1}{m}.$$

Thus, the limit exists and is unique and finite, therefore the given integral is convergent.

Example 29:

Test the convergence of $\int_0^\infty e^{2x}dx,$

Solution:

We have $\int_0^\infty e^{2x}dx = \lim_{x\to\infty}\int_0^x e^{2x}dx,$ (By def.)

$$= \lim_{x\to\infty}\left[\frac{e^{2x}}{2}\right]_0^x = \frac{1}{2}\lim_{x\to\infty}[e^{2x}-1] = \infty.$$

Thus the limit does not exist finitely and therefore the given integral is divergent (*i.e.*, the integral does not exist).

Example 30:

Test the convergence of $\int_0^\infty \frac{4a\,dx}{x^2+4a^2}.$

Solution:

We have $\int_0^\infty \frac{4a\,dx}{x^2+4a^2} = \lim_{x\to\infty}\int_0^x \frac{4a\,dx}{x^2+(2a)^2},$ (By def.)

$$= \lim_{x\to\infty}\left[4a\cdot\frac{1}{2a}\tan^{-1}\frac{x}{2a}\right]_0^x = 2\lim_{x\to\infty}\left[\tan^{-1}\frac{x}{2a}\right]_0^x$$

$$= 2\cdot\lim_{x\to\infty}\left[\tan^{-1}\frac{x}{2a}-0\right] = 2.\ [\tan^{-1}\infty] = 2.\frac{\pi}{2} = \pi.$$

Thus, the limit exists and is unique and finite; therefore, the given integral is convergent.

Example 31:

Test the convergence of $\int_{-\infty}^\infty e^{-x}dx.$

Solution:

We have

$$\int_{-\infty}^{\infty} e^{-x}dx = \int_{-\infty}^{0} e^{-x}dx + \int_{0}^{\infty} e^{-x}dx$$
$$= +\infty + 1 = \infty.j$$

Thus, the limit does not exist finitely. Therefore the given integral is divergent (*i.e.*, the integral does not exist).

Example 32:

Test the convergence of $\int_{-\infty}^{\infty} \frac{dx}{1+x^2}$

Solution:

We have

$$\int_{-\infty}^{\infty} \frac{dx}{1+x^2} = \int_{-\infty}^{\infty} \frac{dx}{1+x^2} + \int_{0}^{\infty} \frac{dx}{1+x^2}$$
$$= \lim_{x\to\infty} \int_{-x}^{0} \frac{dx}{1+x^2} + \lim_{x\to\infty} \int_{0}^{x} \frac{dx}{1+x^2}$$
$$= \lim_{x\to\infty} \left[\tan^{-1} x\right]_{-x}^{0} + \lim_{x\to\infty} \left[\tan^{-1} x\right]_{0}^{x}$$
$$= \lim_{x\to\infty} [0 - \tan^{-1}(-x)] + \lim_{x\to\infty} [\tan^{-1} x - 0]$$
$$= -(-\pi/2) + \pi/2 = \pi.$$

Thus, the limit exists and is unique and finite; therefore the given integral is convergent.

Example 33:

Evaluate $\int_{-\infty}^{\infty} \frac{dx}{x^2+2x+2}$.

Solution:

We have $\int_{-\infty}^{\infty} \frac{dx}{x^2+2x+2} = \int_{-\infty}^{\infty} \frac{dx}{(x+1)^2+1}$

$$= \lim_{x_1\to\infty} \int_{-x_1}^{c} \frac{dx}{(x+1)^2+1} + \lim_{x_2\to\infty} \int_{c}^{x_2} \frac{dx}{(x+1)^2+1},$$

Where c is any real number

$$= \lim_{x_1\to\infty} \left[\tan^{-1}(x+1)\right]_{-x_1}^{c} + \lim_{x_2\to\infty} \left[\tan^{-1}(x+1)\right]_{c}^{x_2}$$
$$= \lim_{x_1\to\infty} [\tan^{-1}(c+1) + \tan^{-1}(1-x_1)]$$
$$+ \lim_{x_2\to\infty} [\tan^{-1}(x_2+1) - \tan^{-1}(c+1)]$$

$$= \tan^{-1}(c+1) - \tan^{-1}(-\infty) + \tan^{-1}(\infty) - \tan^{-1}(c+1)$$

$$= -(-\pi/2) + \pi/2 = \pi.$$

Hence the given integral is convergent and its value is π.

Example 34:

Evaluate $\int_0^1 \frac{dx}{\sqrt{x}}$.

Solution:

In the given integral, the integrand $1/\sqrt{x}$ becomes infinite at the lower limit $x = 0$. Therefore we have

$$\int_0^1 \frac{dx}{\sqrt{x}} = \lim_{\varepsilon \to 0} \int_{0+\varepsilon}^1 \frac{dx}{\sqrt{x}} = \lim_{\varepsilon \to 0} \left[2\sqrt{x}\right]_\varepsilon^1$$

$$= \lim_{\varepsilon \to 0} [2 - 2\sqrt{\varepsilon}] = 2.$$

Hence the given integral is convergent and its value is 2.

Example 35:

Evaluate $\int_0^1 \frac{dx}{x^3}$.

Solution:

Here the integrand (*i.e.*, $1/x^3$) becomes infinite at the lower limit $x = 0$.

$$\therefore \quad \int_0^1 \frac{dx}{x^3} = \lim_{\varepsilon \to 0} \int_{0+\varepsilon}^1 \frac{dx}{x^3} = \lim_{\varepsilon \to 0} \left[-\frac{1}{2x^2}\right]_\varepsilon^1$$

$$= \lim_{\varepsilon \to 0} \left[-\frac{1}{2} + \frac{1}{2\varepsilon^2}\right]$$

$$= -\frac{1}{2} + \infty = \infty.$$

Hence the limit does not exist finitely. Therefore the given integral diverges to ∞ and its value does not exist.

Example 36:

Evaluate $\int_0^1 \frac{dx}{\sqrt{(1-x)}}$.

Solution:

Here the integrand *i.e.*, $1/\sqrt{(1-x)}$ becomes unbounded *i.e.*, infinite at the upper limit (*i.e.*, $x = 1$).

$$\therefore \int_0^1 \frac{dx}{\sqrt{(1-x)}} = \lim_{\varepsilon \to 0} \int_0^{1-\varepsilon} \frac{dx}{\sqrt{(1-x)}}$$

$$= \lim_{\varepsilon \to 0}\left[-2\sqrt{(1-x)}\right]_0^{1-\varepsilon}$$

$$= \lim_{\varepsilon \to 0}\ [-2\sqrt{\varepsilon} + 2] = 2,$$

which is a definite real number. Hence the given integral is convergent and its value is 2.

Example 37:

Evaluate $\int_0^1 \frac{dx}{1-x}$.

Solution:

Here the integrand [*i.e.*, $1/\sqrt{(1-x)}$] becomes infinite at the upper limit (*i.e.*, x = 1).

$$\therefore \int_0^1 \frac{dx}{1-x} = \lim_{\varepsilon \to 0}\int_0^{1-\varepsilon}\frac{dx}{1-x} = \lim_{\varepsilon \to 0}\left[-\log(1-x)\right]_0^{1-\varepsilon}$$

$$= \lim_{\varepsilon \to 0}\ [-\log\varepsilon + 0] = -[-\infty] = \infty.$$

Hence the limit does not exist finitely and therefore the given integral does not exist *i.e.*, is divergent.

Example 38:

Evaluate $\int_{-1}^1 \frac{dx}{x^{2/3}}$.

Solution:

Here the integrand becomes infinite at x = 0 and 0 lies between – 1 and 1.

$$\therefore \int_{-1}^1 \frac{dx}{x^{2/3}} = \int_{-1}^1 \frac{dx}{x^{2/3}} + \int_0^1 \frac{dx}{x^{2/3}}$$

$$= \lim_{\varepsilon \to 0}\int_{-1}^{0-\varepsilon}\frac{dx}{x^{2/3}} + \lim_{\varepsilon' \to 0}\int_{\varepsilon+0}^{1}\frac{dx}{x^{2/3}}$$

$$= \lim_{\varepsilon \to 0}\left[3x^{1/3}\right]_{-1}^{-\varepsilon} + \lim_{\varepsilon' \to 0}\left[3x^{1/3}\right]_{\varepsilon}^{1}$$

$$= \lim_{\varepsilon \to 0}[-3\varepsilon^{1/3}+3] + \lim_{\varepsilon' \to 0}[3-3(\varepsilon')^{1/3}] = 3 + 3 = 6.$$

Example 39:

Evaluate $\int_{-1}^1 \frac{dx}{x^2}$.

Solution:

Here the integrand becomes infinite at x = 0 and $-1 < 0 < 1$.

$$\therefore \int_{-1}^{1} \frac{dx}{x^2} = \lim_{\varepsilon \to 0} \int_{-1}^{-\varepsilon} \frac{dx}{x^2} + \lim_{\varepsilon' \to 0} \int_{\varepsilon'}^{1} \frac{dx}{x^2}$$

$$= \lim_{\varepsilon \to 0} \left[-\frac{1}{x} \right]_{-1}^{-\varepsilon} + \lim_{\varepsilon' \to 0} \left[-\frac{1}{x} \right]_{\varepsilon'}^{1}$$

$$= \lim_{\varepsilon \to 0} \left[\frac{1}{\varepsilon} - 1 \right] + \lim_{\varepsilon' \to 0} \left[-1 + \frac{1}{\varepsilon'} \right].$$

Since both the limits do not exist finitely, therefor the integral does note exist and is divergent.

Example 40:

Evaluate $\int_{0}^{2a} \frac{dx}{(x-a)^2}$.

Solution:

Here the integrand becomes infinite at x = a and 0 < a < 2a. We find that the limits do not exist finitely *i.e.*, the given integral is meaningless.

Example 41:

Test the convergence of the integral

$$\int_{0}^{\infty} \frac{\cos mx}{x^2 + ^2} dx.$$

Solution:

Here $f(x) = \frac{\cos mx}{x^2 + a^2}$. Let $g(x) = \frac{1}{x^2 + a^2}$.

Obviously g (x) is positive in the interval $(0, \infty)$.

we have $|f(x)| = \left| \frac{\cos mx}{x^2 + a^2} \right| = \frac{|\cos mx|}{x^2 + a^2}$

$\leq \frac{1}{x^2 + a^2}$., since $|\cos mx| \leq 1$.

Thus $|f(x)| \leq g(x)$ when $x \geq 0$.

$\therefore$ by comparison test, $\int_{0}^{\infty} \frac{\cos mx}{x^2 + a^2} dx$ is convergent if $\int_{0}^{\infty} \frac{dx}{x^2 + a^2} dx$ is convergent.

But $\int_{0}^{\infty} \frac{dx}{x^2 + a^2} dx = \lim_{x \to \infty} \int_{0}^{x} \frac{dx}{x^2 + s^2} = \lim_{x \to \infty} \left[\frac{1}{a} \tan^{-1} \frac{x}{a} \right]_{0}^{x}$

$= \lim_{x \to \infty} \left[\frac{1}{a} \tan^{-1} \frac{x}{a} - 0 \right] = \frac{1}{a} \cdot \frac{\pi}{2}$ = a definite real number.

$\therefore \int_0^\infty \frac{dx}{x^2 + a^2}$ is convergent.

Hence $\int_0^\infty \frac{\cos mx}{x^2 + a^2} dx$ is also convergent.

Example 42:

Test the convergence of the integral

$$\int_0^\infty \frac{\cos x}{1 + x^2} dx.$$

Solution:

Do your self.

Taking m = 1 and a = 1.

Example 43:

Test the convergence of the integral

$$\int_0^\infty \frac{\sin^2 x}{x^2} dx$$

Solution:

Let a > 0. Then we can write

$$\int_0^\infty \frac{\sin^2 x}{x^2} dx = \int_0^a \frac{\sin^2 x}{x^2} dx + \int_a^\infty \frac{\sin^2 x}{x^2} dx.$$

Since $\lim_{x \to 0} \frac{\sin^2 x}{x^2} = 1$, therefore the integrand $\frac{\sin^2 x}{x^2}$ is bounded throughout the finite interval (0, a).

So $\int_0^a \frac{\sin^2 x}{x^2} dx$ is a proper integral and we need to check the convergence of the integral $\int_0^\infty \frac{\sin^2 x}{x^2} dx$ only.

Here $f(x) = \frac{\sin^2 x}{x^2}$. Take $g(x) = \frac{1}{x^2}$.

Obviously g (x) is positive in the interval (a, ∞).

We have $|f(x)| = \left|\frac{\sin^2 x}{x^2}\right| = \frac{\sin^2 x}{x^2}$

$\leq \frac{1}{x^2}$, since $\sin^2 x \leq 1$.

$\therefore$ by comparison test, $\int_a^\infty \frac{\sin^2 x}{x^2} dx$ is convergent is $\int_a^\infty \frac{dx}{x^2} dx$ is convergent.

But the comparison integral $\int_a^\infty \frac{dx}{x^2}$ is convergent because here n = 2 which is > 1.

$\therefore \int_a^\infty \frac{\sin^2 x}{x^2} dx$ is convergent.

Hence $\int_0^\infty \frac{\sin^2 x}{x^2} dx$ is convergent.

Example 44:

Show that the integral $\int_\pi^\infty \frac{\sin^2 x}{x^2} dx$ *is convergent.*

Solution:

Do your self.

By taking a = π.

Example 45:

Shwo that the integral $\int_0^1 \frac{dx}{x^{1/3}(1+x^2)}$ *is convergent.*

Solution:

In the given integral, the integrand $f(x) = \int_0^1 \frac{dx}{x^{1/3}(1+x^2)}$ is unbounded at the lower limit of integration x = 0.

Take $g(x) = 1/x^{1/3}$.

Then $\lim_{x \to 0} \frac{f(x)}{g(x)} = \lim_{x \to 0} \frac{1}{1+x^2} = 1$, which is finite and non-zero.

$\therefore$ by comparison test

$$\int_0^1 f(x)dx = \text{and} \int_0^1 g(x)dx$$

either both converge or both diverge. But the comparison integral $\int_0^1 \frac{dx}{x^{1/3}}$ is convergent because here n = 1/3 which is less than is less than 1. Hence the integral $\int_0^1 \frac{dx}{x^{1/3}(1+x^2)}$ is also convergent.

Example 46:

Test the convergence of the integral

$$\int_{\pi}^{\infty} \frac{\sin x}{x^2} dx.$$

Solution:

Let $f(x) = \frac{\sin x}{x^2}$. Then g (x) is positive in the interval (π, ∞). Take (x) = $1/x^2$. Then g(x) is positive in the interval (p, ∞).

We have $|f(x)| = \left|\frac{\sin x}{x^2}\right| = \frac{|\sin x|}{x^2}$

$\leq \frac{1}{x^2}$, since $|\sin x| \leq 1$.

$\therefore$ by comparison test, $\int_{\pi}^{\infty} \frac{\sin x}{x^2} dx$ is convergent is $\int_{\pi}^{\infty} \frac{dx}{x^2}$ is convergent.

But the comparison integral $\int_{\pi}^{\infty} \frac{dx}{x^2}$ is convergent because here n = 2 which is > 1.

Hence $\int_{\pi}^{\infty} \frac{\sin x}{x^2} dx$ is also convergent.

Example 47:

Show that the integral $\int_{a}^{\infty} \frac{dx}{x\sqrt{(1+x^2)}}$ *converges, where a > 0.*

Solution:

Let $f(x) = \frac{1}{x\sqrt{(1+x^2)}}$.

Then f (x) is bounded in the interval (a, ∞). Take g (x) = $1/x^2$.

Then g (x) is positive in the interval (a, ∞). We have

$$|f(x)| = \left|\frac{1}{x\sqrt{(1+x^2)}}\right| = \frac{x}{x^2\sqrt{\{1+(1/x^2)\}}}$$

$< \frac{1}{x^2}$, since $\frac{1}{\sqrt{\{1+(1/x^2)\}}} < 1$.

$\therefore$ by the comparison integral $\int_{a}^{\infty} \frac{dx}{x\sqrt{(1+x^2)}}$ is convergent if $\int_{a}^{\infty} \frac{dx}{x^2}$ is convergent.

But comparison integral $\int_{a}^{\infty} \frac{dx}{x^2}$ is convergent because here n = 2 which is > 1.

Hence $\int_{a}^{\infty} \frac{dx}{x\sqrt{(1+x^2)}}$ is also convergent.

Alternative Method

Here $f(x) = \dfrac{1}{x\sqrt{(1+x^2)}} = \dfrac{1}{x^2\sqrt{\{1+(1/x^2)\}}}$. Take $g(x) = \dfrac{1}{x^2}$.

We have $\lim_{x \to \infty} \dfrac{f(x)}{g(x)} = \lim_{x \to \infty} \dfrac{1}{\sqrt{\{1+(1/x^2)\}}} = 1$, which is finite and non-zero therefore $\int_a^{\infty} f(x)dx$ and $\int_a^{\infty} g(x)dx$ either both converge or both diverge. But $\int_a^{\infty} g(x)dx = \int_a^{\infty} \dfrac{dx}{x^2}$ is convergent because here n = 2 which is > 1.

Hence $\int_a^{\infty} f(x)dx$ *i.e.*, $\int_a^{\infty} \dfrac{1}{x\sqrt{(1+x^2)}}dx$ is also convergent.

Example 48:

Test the convergent of $\int_0^2 \dfrac{\log x}{\sqrt{(2-x)}} dx$.

Solution:

Let $f(x) = \dfrac{\log x}{\sqrt{(2-x)}}dx$. Then f(x) is unbounded both at x = 0 and x = 2. If 0 < a < 2, we can write

$$\int_0^2 \frac{\log x}{\sqrt{(2-x)}}dx = \int_0^a \frac{\log x}{\sqrt{(2-x)}}dx + \int_a^2 \frac{\log x}{\sqrt{(2-x)}}dx$$

$= I_1 + I_2$, say.

To test the convergence of I_1. We have

$$\lim_{x \to 0} x^{\mu} f(x) = \lim_{x \to 0}\left\{x^{\mu} \cdot \frac{\log x}{\sqrt{(x-2)}}\right\} = 0 \text{ if } \mu > 0.$$

Therefore taking μ between 0 and 1, it follows by μ-test that I_1 is convergent.

To test the convergence of I_2. Take $\mu = \dfrac{1}{2}$. We have

$$\lim_{x \to 2-0} (2-x)^{\mu}.\ f(x) = \lim_{x \to 2-0} (2-x)^{1/2} \cdot \frac{\log x}{\sqrt{(2-x)}}$$

$$= \lim_{x \to 2-0} \log x = \lim_{\varepsilon \to 0} \log(2-\varepsilon) = \log 2.$$

∴ by μ-test I_2 is convergent because 0 < μ < 1.

Hence, the given integral is also convergent, it being the sum of two convergent integrals.

Example 49:

Test the convergence of $\int_0^1 x^{p-1}e^{-x}dx$.

Solution:

Let $f(x) = x^{p-1}e^{-x}$ and $I = \int_0^1 x^{p-1}e^{-x}dx$.

If $p \geq 1$, f(x) is bounded throughout the interval (0, 1) and so I is a proper integral and hence it is convergent if $p \geq 1$.

If $p < 1$, f(x) is unbounded at x = 0. In this case, we have

$$\lim_{x \to 0} x^{\mu} f(x) = \lim_{x \to 0} x^{\mu}.x^{p-1}e^{-x} = \lim_{x \to 0} x^{\mu+p-1}e^{-x}$$

$= 1$ is $\mu + p - 1 = 0$ *i.e.*, $\mu = 1 - p$.

So by μ-test when $0 < \mu < 1$ *i.e.*, $0 < p < 1$, the given integral is convergent and when $\mu \geq 1$ *i.e.*, $p \leq 0$, the given integral is divergent.

Hence I is convergent if $p > 0$ and is divergent if $p \leq 0$.

Example 50:

Test the convergence of the integral $\int_0^1 \frac{dx}{x^{1/3}(1+x^2)}$.

Solution:

In the given integral the integrand $f(x) = \frac{dx}{x^3(1+x^2)}$ is unbounded at the lower limit of integration x = 0. Take $g(x) = 1/x^3$.

Then $$\lim_{x \to 0} \frac{f(x)}{g(x)} = \lim_{x \to 0} \frac{1}{1+x^2} = 1,$$

which is finite and non-zero. Therefore, by comparison test,

$$\int_0^1 f(x)dx \text{ and } \int_0^1 g(x)dx$$

either both converge or both diverge. But the comparison integral $\int_0^1 \frac{dx}{x^3}$ is divergent because here n = 3 which is > 1.

Hence the given integral $\int_0^1 \frac{dx}{x^3(1+x^2)}$ is also divergent.

Example 51:

Test the convergence of the integral $\int_1^2 \frac{dx}{\sqrt{(x^4-1)}}$.

Solution:

In the given integral the integrand $f(x) = 1/\sqrt{(x^4 - 1)}$ is anbounded at the lower limit of integration x = 1.

Take $g(x) = 1/\sqrt{(x^2 - 1)}$.

Then $\lim_{x \to 1} \frac{f(x)}{g(x)} = \lim_{x \to 1} \left\{ \frac{1}{\sqrt{(x^4 - 1)}} \cdot \sqrt{(x^2 - 1)} \right\}$

$= \lim_{x \to 1} \frac{1}{\sqrt{(x^2 + 1)}}$

$= 1/\sqrt{2}$, which is finite and non-zero.

Therefore by comparison test,

$\int_1^2 f(x)dx$ and $\int_1^2 g(x)dx$

are either both convergent or both divergent.

But $\int_1^2 g(x)dx = \int_1^2 \frac{dx}{\sqrt{(x^2 - 1)}} = \lim_{\varepsilon \to 0} \int_{1+\varepsilon}^2 \frac{dx}{\sqrt{(x^2 - 1)}}$

$= \lim_{\varepsilon \to 0} \left[\log\{x + \sqrt{(x^2 - 1)}\} \right]_{1+\varepsilon}^2$

$= \lim_{\varepsilon \to 0} [\log(2 + \sqrt{3}) - \log\{1 + \varepsilon + \sqrt{(\varepsilon^2 + \varepsilon)}\}] = \log(2 + \sqrt{3})$, which is a definite real number.

$\therefore \int_1^2 g(x)dx$ is convergent.

Hence $\int_1^2 \frac{1}{(x^4 - 1)} dx$ is also convergent.

Example 52:

Test the convergence of the integral

$$\int_0^1 \frac{dx}{(x+1)\sqrt{(1-x^2)}}.$$

Solution:

In the given integral the integrand $f(x) = \frac{1}{(x+1)\sqrt{(1-x^2)}}$ is unbounded at the upper limit of integration $x = 1$. Take

$$g(x) = \frac{1}{\sqrt{(1-x^2)}}$$

Then $\lim_{x \to 1} \frac{f(x)}{g(x)} = \lim_{x \to 1} \frac{1}{x+1} = \frac{1}{2}$.

Which is finite and non-zero.

Therefore, by comparison test.

$$\int_0^1 f(x)dx \text{ and } \int_0^1 g(x)dx$$

either both converge or both diverge.

But $\int_0^1 g(x)dx = \int_0^1 \frac{dx}{\sqrt{(1-x^2)}} = \lim_{\varepsilon \to 0} \int_0^{1-\varepsilon} \frac{dx}{\sqrt{(1-x^2)}}$

$= \lim_{\varepsilon \to 0} \left[\sin^{-1} x\right]_0^1 = \lim_{\varepsilon \to 0}$ [$\sin^{-1}(1-\varepsilon) = \sin^{-1} 1 = \pi/2$.

which is a definite real number.

$\therefore \int_0^1 g(x)dx$ is convergent.

Hence $\int_0^1 \frac{dx}{(x+1)\sqrt{(1-x^2)}}$ is also convergent.

Example 53:

Show that the integral $\int_0^1 \frac{\sec x}{x} dx$ *is divergent.*

Solution:

In the given integral the integrand $f(x) = \int_0^1 \frac{\sec x}{x} dx$ is unbounded at the lower limit of integration $x = 0$. Take $g(x) = 1/x$.

Then $\lim_{x \to 0} \frac{f(x)}{g(x)} = \lim_{x \to 0} \left\{\frac{\sec x}{x} \cdot x\right\} = \lim_{x \to 0} \sec x = 1$,

which is finite and non-zero.

Therefore, by comparison test,

$$\int_0^1 f(x)dx \text{ and } \int_0^1 g(x)dx$$

either both converge or both diverge. But the comparison integral $\int_0^1 \frac{1}{x} dx$ is divergent because here $n = 1$.

Hence the given integral $\int_0^1 \frac{\sec x}{x} dx$ is also divergent.

Example 54:

Test the convergence of the integral $\int_0^{\pi/2} \frac{\cos x}{x^2} dx$.

Solution:

In the given integral the integrand $f(x) = \frac{\cos x}{x^2}$ is unbounded at the lower limit of integration $x = 0$. Take $g(x) = 1/x^2$.

Then $\lim_{x \to 0} \frac{f(x)}{g(x)} = \lim_{x \to 0} \left\{ \frac{\cos x}{x^2} \cdot x^2 \right\} = \lim_{x \to 0} \cos x = 1,$ which is finite and non-zero.

Therefore, by comparison test, $\int_0^{\pi/2} f(x)dx$ and $\int_0^{\pi/2} g(x)dx$ either both converge or both diverge.

But $\int_0^{\pi/2} g(x)dx = \int_0^{\pi/2} \frac{1}{x^2} dx = \lim_{\varepsilon \to 0} \int_\varepsilon^{\pi/2} \frac{1}{x^2} dx$

$$= \lim_{\varepsilon \to 0} \left[-\frac{1}{x} \right]_\varepsilon^{\pi/2} = \lim_{\varepsilon \to 0} \left[-\frac{2}{\pi} + \frac{1}{\varepsilon} \right] = \infty.$$

$\therefore \int_0^{\pi/2} g(x)dx$ is divergent.

Hence the given integral $\int_0^{\pi/2} \frac{\cos x}{x^2} dx$ is also divergent.

Example 55:

Show that $\int_0^1 x^{n-1} e^{x} dx$ *is convergent if* $n > 0$.

Solution:

If $n \geq 1$, then $\int_0^1 x^{n-1} e^{-x} dx$ is a proper integral because the integrand $f(x) = x^{n-1} e^{-x}$ is bounded in the interval (0, 1). So the given integral is convergent when $n \leq 1$.

If $0 < n < 1$, the integrand $f(x) = x^{n-1}e^{-x}$ is unbounded at $x = 0$.

Take $g(x) = x^{n-1}$.

Then $\lim_{x \to 0} \frac{f(x)}{g(x)} = \lim_{x \to 0} \frac{f(x)}{g(x)} e^{-x} = 1$, which is finite and non-zero.

$\therefore$ by comparison test, $\int_0^1 f(x)dx$ and $\int_0^1 g(x)dx$ either both converge or both diverge.

But $\int_0^1 g(x)dx = \int_0^1 x^{n-1}dx = \lim_{\varepsilon \to 0} \int_\varepsilon^1 x^{n-1}dx = \lim_{\varepsilon \to 0} \left[\frac{x^n}{n} \right]_\varepsilon^1$

$= \lim_{\varepsilon \to 0} \left[\frac{1}{n} - \frac{\varepsilon^n}{n} \right] = \frac{1}{n}$, which is a definite real number.

$\therefore \int_0^1 g(x)dx$ is convergent.

Hence $\int_0^1 x^{n-1}e^{-x}dx$ is also convergent.

Example 56:

Show that the integral $\int_0^\infty x^{n-1}e^{-x}dx$ *is convergent if* $n > 0$.

Solution:

We have $\int_0^\infty x^{n-1}e^{-x}dx = \int_0^1 x^{n-1}e^{-x}dx + \int_1^\infty x^{n-1}e^{-x}dx.$

Let $I_1 = \int_0^1 x^{n-1}e^{-x}dx$ and $I_2 = \int_1^\infty x^{n-1}e^{-x}dx.$

The integral I_2 is convergent for all vaues of n.

Also the integral I_1 is convergent if $n > 0$.

Hence the given integral is convergent if $n > 0$ because then it is the sum of two convergent integrals.

Example 57:

Prove that the integral $\int_0^1 \frac{dx}{\sqrt{\{x(1-x)\}}}$ *converges.*

Solution:

In the given integral the integrand $f(x) = 1/\sqrt{\{x(1-x)\}}$ is unbounded both at $x = 0$ and at $x = 1$. If $0 < a < 1$, we can write

$$\int_0^1 \frac{dx}{\sqrt{\{x(1-x)\}}} = \int_0^a \frac{dx}{\sqrt{\{x(1-x)\}}} + \int_a^1 \frac{dx}{\sqrt{\{x(1-x)\}}}$$

$= I_1 + I_2$, say.

In the integral I_1 the integrand $f(x)$ is unbounded at the lower limit of integration $x = 0$ and in the integral I_2 the integrand $f(x)$ is unbounded at the upper limit of integration $x = 1$.

To test the convergence of I_1. Take $\mu = \frac{1}{2}$. We have

$$\lim_{x \to 0} x^\mu f(x) = \lim_{x \to 0} x^{1/2} \cdot \frac{1}{\sqrt{\{x(1-x)\}}} = \lim_{x \to 0} \frac{1}{\sqrt{(1-x)}}$$

$= 1$ *i.e.*, the limit exists.

Since $0 < \mu < \frac{1}{2}$, therefore by μ-test I_1 is convergent.

To test the convergence of I_2. Take $\mu = \frac{1}{2}$. We have

$$\lim_{x \to 1-0} (1-x)^\mu . f(x) = \lim_{x \to 1-0} (1-x)^{1/2} \cdot \frac{1}{\sqrt{\{x(1-x)\}}}$$

$$= \lim_{x \to 1-0} \frac{1}{\sqrt{x}} = \lim_{\varepsilon \to 0} \frac{1}{\sqrt{(1-\varepsilon)}} = 1.$$

Hence by μ-test I_2 is convergent since $0 < \mu < 1$.

Thus the given integral is the sum of two converged integrals.

Hence the given integral itself is convergent.

Example 58:

Test the convergence of $\int_0^{\pi/4} \frac{1}{\sqrt{(tan\, x)}}\, dx$.

Solution:

Here the integrand $f(x) = 1\ \sqrt{(\tan x)}$ is unbounded at $x = 0$.

Take $\mu = \frac{1}{2}$.

We have $\lim\limits_{x \to 0} x^{\mu}.f(x) = \lim\limits_{x \to 0} x^{1/2} \cdot \frac{1}{\sqrt{(\tan x)}}$

$= \lim\limits_{x \to 0} \sqrt{\left(\frac{x}{\sin x}\right)} \cdot \sqrt{(\cos x)} = 1.1 = 1.$

Since $0 < \mu < 1$, therefore by μ-test the given integral is convergent.

Example 59:

Test the convergence of $\int_0^{\pi/2} \frac{\sin x}{x^{1+n}}\, dx$.

Solution:

We have $\int_0^{\pi/2} \frac{\sin x}{x^{1+n}}\, dx = \int_0^{\pi/2} \left(\frac{\sin x}{x}\right) \cdot \frac{1}{x^n}\, dx.$

Now $\lim\limits_{x \to 0} \frac{\sin x}{x} = 1$. Therefore, for $n \leq 0$, the integrand is bounded throughout the interval $(0, \pi/2)$ and so the given integral is a proper integral and hence it is convergent if $n \leq 0$.

If $n > 0$, the integrand is unbounded only at $x = 0$. In this case, we have

$$\lim_{x \to 0} x^{\mu} \frac{\sin x}{x^{n+1}} = \lim_{x \to 0} \left\{\left(\frac{\sin x}{x}\right) \cdot \frac{x^{\mu}}{x^{n}}\right\}$$

$$= \lim_{x \to 0} \left\{x^{\mu - n} \cdot \left(\frac{\sin x}{x}\right)\right\} = 1, \text{ if } \mu - n\ 0\ \textit{i.e.}, \mu = n.$$

∴ by μ-test if $0 < \mu < 1$ *i.e.*, $0 < n < 1$, the given integral is convergent and if $\mu \geq 1$ *i.e.*, $n \geq 1$, the given integral is divergent.

Hence the given integral is convergent if $n < 1$ and divergent if $n \geq 1$.

Example 60:

Test the convergence of $\int_0^{\pi/2} \frac{\cos x}{x^n} dx$.

Solution:

When $n \le 0$, the given integral is a proper integral and hence convergent.

When $n > 0$, the integrand becomes unbounded at $x = 0$.

Let $f(x) = \frac{\cos x}{x^n}$.

Then $\lim_{x \to 0} x^\mu f(x) = \lim_{x \to 0} x^{\mu - n} \cos x = 1$, if $\mu = n$.

Hence by μ-test it follows that the given integral is convergent when $0 < n < 1$, and divergent when $n \ge 1$.

From the above discussion we conclude that the given integral is convergnet when $n < 1$, and divergent when $n \ge 1$.

Example 61:

Show that the integral $\int_0^{\pi/2} \log \sin x \, dx$ *converges.*

Solution:

The only point of infinite discontinuity of the integrand is $x = 0$.

Now $\lim_{x \to 0} x^\mu \log \sin x$, when $m > 0$

$$= \lim_{x \to 0} \frac{\log \sin x}{x^{-\mu}}, \qquad \left[\text{form } \frac{\infty}{\infty}\right]$$

$$= \lim_{x \to 0} \frac{\cos x}{-\mu x^{-\mu-1}}$$

$$= \lim_{x \to 0} -\frac{1}{\mu} \cdot \frac{x^{\mu+1}}{\tan x} \qquad \left[\text{form } \frac{0}{0}\right]$$

$$= \lim_{x \to 0} -\frac{1}{\mu} \cdot \frac{(\mu+1)}{\sec^2 x}, \qquad \text{[by L-Hospital's rule]}$$

$= 0$, if $\mu > 0$.

Taking μ between 0 and 1, it follows from μ-test that the given integral is convergent.

Example 62:

Discuss the convergence of the integral

$\int_0^1 x^{n-1} \log x \, dx$.

Solution:

(i) Since $\lim_{x \to 0} x^r \log x = 0$ where $r > 0$, the integral is a proper integral, when $n > 1$.

(ii) When $n = 1$, we have

$$\int_0^1 \log x \, dx = \lim_{\varepsilon \to 0} \int_\varepsilon^1 \log x \, dx$$

$$= \lim_{\varepsilon \to 0} [x \log x - x]_\varepsilon^1$$

$$= \lim_{\varepsilon \to 0} [-1 - \varepsilon \log \varepsilon + \varepsilon] = -1.$$

$\therefore$ the integral is convergent if $n = 1$.

(iii) Let $n < 1$ and $f(x) = x^{n-1} \log x$.

Then $\lim_{\varepsilon \to 0} x^\mu f(x) = \lim_{x \to 0} x^{\mu+n-1} \log x$

$= 0$ if $\mu > 1 - n$ ...(1)

and $= -\infty$ if $\mu \leq 1 - n$. ...(2)

Hence, when $0 < n < 1$, we can choose μ between 0 and 1 and satisfying (1). The integral is therefore convergent by μ-test when $0 < n < 1$.

Again when $n \leq 0$, we can take $\mu = 1$ and satisfying (2). Hence, by μ-test the integral is divergent when $n \leq 0$.

Therefore, from (i), (ii) and (iii), we conclude that the given integral is convergent when $n > 0$ and divergent when $n \leq 0$.

Example 63:

Show that the integral

$$\int_0^\infty e^{-a^2x^2} \cos bx \, dx$$

is absolutely convergent.

Solution:

We have

$$\int_0^\infty \left| e^{-a^2x^2} \cos bx \right| dx \leq \int_0^\infty \left| e^{-a^2x^2} \right| dx$$

$$= \int_0^\infty e^{-a^2x^2} dx$$

$$= \int_0^c e^{-a^2x^2}\,dx + \int_c^\infty e^{-a^2x^2}\,dx \text{ where } c > 0.$$

But the integral $\int_0^c e^{-a^2x^2}\,dx$ is a proper integral and hence convergent.

Also $\int_c^\infty e^{-a^2x^2}\,dx$ is convergent by μ-test, for we have

$$\lim_{x\to\infty} x^\mu e^{-a^2x^2} = \lim_{x\to\infty} \frac{x^\mu}{1 + a^2x^2 + \dfrac{a^4x^4}{2!} + \ldots}$$

$= 0$, for all values of μ.

Taking $\mu > 1$, we see that $\int_c^\infty e^{-a^2x^2}\,dx$ is convergent.

$\therefore \int_0^\infty e^{-a^2x^2}\,dx$ is convergent.

Hence $\int_0^\infty \left|e^{-a^2x^2} \cos bx\right| dx$ is convergent, by comparison test.

Form the above discussion it follows that the given integral is absolutely convergent.

Example 64:

Discuss the convergence or divergence of the integral

$$\int_0^\infty \frac{x^{a-1}}{1+x}\,dx.$$

Solution:

Let $f(x) = \dfrac{x^{a-1}}{1+x}$. If $b > 0$, we can write

$$\int_0^\infty \frac{x^{a-1}}{1+x}\,dx = \int_0^b \frac{x^{a-1}}{1+x}\,dx + \int_b^\infty \frac{x^{a-1}}{1+x}\,dx$$

$= I_1 + I_2$, say.

Let $a \geq 1$. Then $f(x)$ is bounded throughout the interval $(0, b)$ and so the integral I_1 is a proper integral and hence it is convergent. To test the convergence of the infinite integral I_2 in this case, we have

$$\lim_{x\to\infty} x^\mu f(x)$$

$$= \lim_{x\to\infty} x^\mu \cdot \frac{x^{a-1}}{1+x}$$

$$= \lim_{x\to\infty} \frac{x^{\mu+a-1}}{x+1}$$

$= 1$, if $\mu + a - 1 = 1$

i.e., if $\mu = 2 - a$ which is ≤ 1 since $a \geq 1$.

Hence by μ-test I_2 is divergent.

$\therefore$ the given integral is divergent if $a \geq 1$.

Let $a < 1$. Then in the interval (0, b), f (x) is unbounded only at $x = 0$. Also f (x) is bounded throughout the interval (b, ∞). Therefore, in this case I_1 is an improper integral of the second king and I_2 is an improper integral of the first kind. To test the convergence of I_1, we have

$$\lim_{x \to 0} x^{\mu} \cdot \frac{x^{a-1}}{x+1}$$

$$= \lim_{x \to 0} \frac{x^{\mu+a-1}}{x+1}$$

$= 1$ is $\mu + a - 1 = 0$

i.e., if $\mu = 1 - a$.

If we take $0 < a < 1$, then have $0 < \mu < 1$ and so by μ-test I_1 is convergent. If we take $a \leq 0$, then $\mu \geq 1$ and so by μ-test I_1 is divergent.

To test the convergence of I_2 when $a < 1$, we have

$$\lim_{x \to \infty} x^{\mu} \cdot \frac{x^{a-1}}{x+1}$$

$$= \lim_{x \to \infty} \frac{x^{\mu+a+1}}{x+1} = 1,$$

if $\mu + a - 1 = 1$

i.e., if $\mu = 2 - a$ which is > 1 since $a < 1$.

Hence by μ-test I_2 is convergent if $a < 1$.

Thus, I_2 is convergent if $a < 1$. But I_1 is convergent if $0 < a < 1$ and is divergent if $a \leq 0$.

$\therefore$ the given integral is convergent if $0 < a < 1$ and is divergent if $a \leq 0$.

Hence the given integral is convergent if $0 < a < 1$ and is divergent if $a \geq 1$ or if $a \leq 0$.

Example 65:

Discuss the convergence of the Beta function

$$\int_0^1 x^{m-1}(1-x)^{n-1}\,dx.$$

Solution:

Let $f(x) = x^{m-1}(1 - x)^{n-1}$.

Example 66:

Show that $\int_0^{\infty} \frac{dx}{(1+x)^{2/3}}$ *is not convergent.*

Solution:

We have $\int_0^{\infty} \frac{dx}{(1+x)^{2/3}} = \lim_{x \to \infty} \int_0^{x} (1+x)^{-2/3}\,dx$, (By def.)

$$= \lim_{x \to \infty} \left[\frac{(1+x)^{1/3}}{\frac{1}{3}} \right]_0^x = \lim_{x \to \infty} 3\,[(1+x)^{1/3} - 1] = \infty.$$

Thus, the limit does not exist finitely and therefore the given integral is divergent (*i.e.*, the integral does not exist).

Example 67:

Test the convergence of

(i) $\int_{-\infty}^{0} e^{x}dx$; (ii) $\int_{-\infty}^{0} e^{-x}dx$.

Solution:

(i) We have $\int_{-\infty}^{0} e^{x}dx = \lim_{x \to \infty} \int_{-x}^{0} e^{x}dx$, (By def.)

$$= \lim_{x \to \infty} \left[e^{x} \right]_{-x}^{0} = \lim_{x \to \infty} [1 - e^{-x}] = [1 - 0] = 1.$$

Thus, the limits exist and is unique and finite; therefore the given integral is convergent.

(ii) We have $\int_{-\infty}^{0} e^{-x}dx = \lim_{x \to \infty} \int_{-x}^{0} e^{-x}dx$, (By def.)

$$= \lim_{x \to \infty} \left[\frac{e^{-x}}{-1} \right]_{-x}^{0}$$

$$= -\lim_{x \to \infty} [e^{0} - e^{x}] = \infty.$$

Thus, the limit does not exist finitely and therefore the given integral is divergent (*i.e.*, the integral does not exist).

Example 68:

Test the convergence of the following integrals by evaluating them

(i) $\int_{-\infty}^{0} \sinh x\,dx$; (ii) $\int_{-\infty}^{0} \cosh x\,dx$.

Solution:

(i) We have $\int_{-\infty}^{0} \sinh x\, dx = \lim_{x \to \infty} \int_{-x}^{0} \sinh x\, dx$, (By def.)

$$= \lim_{x \to \infty} \int_{-x}^{0} \frac{e^{x} - e^{-x}}{2} dx$$

$$= \frac{1}{2}\left[\lim_{x \to \infty} \int_{-x}^{0} e^{x} dx - \lim_{x \to \infty} \int_{-x}^{0} e^{-x} dx\right]$$

$$= \frac{1}{2}\left[\lim_{x \to \infty} \left\{e^{x}\right\}_{-x}^{0} - \lim_{x \to \infty} \left\{\frac{e^{-x}}{-1}\right\}_{x}^{0}\right]$$

$$= \frac{1}{2}\left[\lim_{x \to \infty} \{e^{0} - e^{-x}\} + \lim_{x \to \infty} \{e^{0} - e^{x}\}\right] = \frac{1}{2}[1 - \infty] = -\infty.$$

Thus, the given integral diverges to $-\infty$.

(ii) We have

$$\int_{-\infty}^{0} \cosh x\, dx = \int_{-\infty}^{0} \frac{e^{x} + e^{-x}}{2} dx$$

$$= \frac{1}{2}\left[\int_{-\infty}^{0} e^{x} dx + \int_{-\infty}^{0} e^{-x} dx\right] = \frac{1}{2}[1 + \infty] = \infty.$$

Thus the given integral diverges to ∞.